Randomization, Bootstrap and
Monte Carlo Methods in Biology,
Second Edition
B.F.J. Manly

Readings in Decision Analysis
S. French

Sampling Methodologies with
Applications
P. Rao

Statistical Analysis of Reliability Data
M.J. Crowder, A.C. Kimber,
T.J. Sweeting and R.L. Smith

Statistical Methods for SPC and TQM
D. Bissell

Statistical Methods in Agriculture and
Experimental Biology, Second Edition
R. Mead, R.N. Curnow and A.M. Hasted

Statistical Process Control — Theory and
Practice, Third Edition
G.B. Wetherill and D.W. Brown

Statistical Theory, Fourth Edition
B.W. Lindgren

Statistics for Accountants, Fourth Edition
S. Letchford

Statistics for Technology —
A Course in Applied Statistics,
Third Edition
C. Chatfield

Statistics in Engineering —
A Practical Approach
A.V. Metcalfe

Statistics in Research and Development,
Second Edition
R. Caulcutt

The Theory of Linear Models
B. Jørgensen

Essential

STATISTICS

FOURTH EDITION

D. G. REES

Formerly Principal Lecturer in Statistics
Oxford Brookes University
UK

CHAPMAN & HALL/CRC

Boca Raton   London   New York   Washington, D.C.

Library of Congress Cataloging-in-Publication Data

Rees, D. G.
    Essential statistics / D.G. Rees.—4th ed.
        p. cm. — (Chapman & Hall texts in statistical science series)
    Includes bibliographical references and index.
    ISBN 1-58488-007-4 (alk. paper)
    1. Statistics. I. Title. II. Texts in statistical science.
QA276.12 R44 2000
519.5—dc21

00-050844
CIP

No claim to original U.S. Government works
International Standard Book Number 1-58488-007-4
Library of Congress Card Number 00-050844
Printed in the United States of America 1 2 3 4 5 6 7 8 9 0
Printed on acid-free paper

**Books are to be returned on or before
the last date below.**

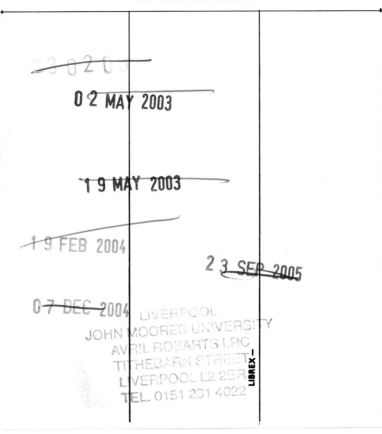

CHAPMAN & HALL/CRC
Texts in Statistical Science Series

Series Editors
C. Chatfield, *University of Bath, UK*
J. Zidek, *University of British Columbia, Canada*

Contents

Preface

This new edition is the result of a careful and thorough review of the third edition. It includes a completely new chapter in which analysis of variance (ANOVA) is introduced. Not only is ANOVA the central technique in the analysis of data from properly designed experiments, but it is also a good way of introducing students to another very important statistical technique, namely, regression analysis.

Because of the limitations of space, only the simplest type of experimental design and analysis is discussed (in Chapter 12). Similarly, only the simplest form of regression analysis is covered (in Chapter 15). Since ANOVA can be extended to the analysis of data from a wide range of experimental designs and also to multiple regression analysis, this introduction could provide a basis for those students who wish to take statistics further, i.e., beyond the scope of this book.

In addition, there is an educational reason for introducing ANOVA; it enables students to see connections (which might otherwise not be appreciated) between two or more statistical concepts. It is hoped that, in this way, the student will begin to see statistics, not as a set of independent topics, but as an integrated subject.

In fact, the more you study statistics, the more I think you will agree with the authors of the book, *Elements of Statistics*,[*] in which they refer to the subject of statistics as "The Queen of the Sciences".

The other main change in this new edition is the replacement of the 'Command version' of Minitab by the 'Windows version'. Those of us, and I include myself, who were first introduced to the former version may have been reluctant to make the change. However, I am convinced that the latter is much better for those students who may already have had some Windows experience when they begin to study statistics (skill with Windows is what educationalists call 'transferable').

---

[*] *Elements of Statistics*, 1st ed., Daly, F., Hand, D.J., Jones, M.C., Lunn, A.D., and McConway, K.J., published for The Open University course M246 by Addison-Wesley-Longman, Reading, MA, 1995. ISBN 0-201-42278-6.

Reverting to the subject of the use of computer packages in statistics, the 'health warning', given in the second paragraph of the Preface to the third edition* still applies; perhaps to an even greater extent, as computer packages become more and more user-friendly. I suggest that you, the student, should read or reread the health warning.

Some of the material from the third edition has been rearranged in this new edition. For example, instead of a multiple choice test at the end of the book, many of these questions have been transferred to the end-of-chapter worksheets. In some worksheets, there is a new final question which is either (a) open-ended, so there is no unique correct answer, or (b) significantly more challenging than the other questions, or (c) one which may require the student to design a project, collect some statistical data (perhaps using e-mail), or (d) do some research using libraries or the Internet.

Appendix A is a new one in that it gives a list of all the statistical formulae used in the book, together with brief notes (where it was thought prudent), explaining in what situation each might be used. Teachers may find that Appendix A could form a set of examination formulae sheets which the student could take into a statistics examination.

Appendix B gives, as in previous editions, detailed solutions to virtually all of the worksheet questions, while Appendices C and D supply a set of statistical tables (including two new ones) and a glossary of symbols, respectively.

Finally, Appendix E provides a brief introduction to Minitab for Windows for students with little or no Windows experience.

I hope that some of my love of statistics has or will come across to you, the student. Let me know if it does.

D.G. REES
e-mail rees@oxfree.com

---

* See the Preface to the third edition on page xv.

Preface to the Third Edition

For this edition, the second edition has been completely reviewed and appropriately revised and rewritten. In addition, there are a number of new sections. For example, Minitab applications have been included within each chapter as they arise, rather than in a special chapter at the end of the book. A short introduction to Minitab is given in an appendix. A new data set (40 cases, 6 variables) has been introduced as a basis for many of the examples in the text. There are new sections on Venn diagrams, the F test for the equality of two variances, the Fisher exact text, the χ^2 trend test and the Shapiro–Wilk test for normality. Some methods applicable to grouped data, for example the mean and standard deviation, have been omitted in this new edition, since it can be assumed that, nowadays, all data are initially input case by case to a computer or calculator. The worksheets at the end of each chapter have also been reviewed and revised. Detailed solutions are again provided, and there is a completely new multiple-choice test.

There is a view that the advent of the statistical computer package has dispensed with the need for the calculator, statistical formulae and statistical tables. I do not share this view. I believe that, for a proper and deep understanding of the concepts of statistics and the analysis of statistical data, it is essential to know what the computer or calculator is doing with the data, what assumptions are being made in carrying out an analysis and whether these assumptions are reasonable assumptions, and also the limitations of each method. The computer may take some of the drudgery out of the calculations, but it is not a substitute for careful thought. The reader will find that virtually all the methods described in this book may be performed by hand, i.e., with a calculator using given formulae and tables, and also by computer, i.e., using Minitab. The underlying assumptions and limitations are given and fully discussed.

Finally, I hope that the friends of *Essential Statistics* who have found earlier editions of value will also like this new edition.

Preface to the Second Edition

The main feature of this new edition is a substantial addition on applications of the interactive statistical computer package, Minitab. This package has become widely used in colleges as an aid to teaching statistics. The new chapter contains over 20 sample programs illustrating how Minitab can be used to draw graphs, calculate statistics, carry out tests and perform simulations. The chapter could act as a primer for first-time Minitab users.

There are also new sections in Chapter 3 and 4 on some aspects of exploratory data analysis. Some changes have been made to the statistical tables. For example, Tables D.1 and D.2 now give cumulative probabilities in terms of 'r or fewer...' instead of 'r or more....' The tables are now consistent with those adopted by most GCSE examination boards and also with the output from the Minitab CDF command for both the binomial and Poisson distributions. For similar reasons Table D.3(a) now gives the cumulative distribution function for normal distribution, i.e., areas to the left of various values of z. Another change is that the conditions for the use of the normal approximation to the binomial have been brought into line with accepted practice. There are other minor changes too numerous to list here.

I am grateful for the opportunity to update and enhance the successful first edition. Many thanks to all those who have expressed their appreciation of *Essential Statistics* as a course text or who have made helpful suggestions for improvements.

Preface to the First Edition

TO THE STUDENT

Are you a student who requires a basic statistics text-book? Are you studying statistics as part of a study of another subject, for example one of the natural, applied or social sciences, or a vocational subject? Do you have an O-level or GCSE in mathematics or an equivalent qualification? If you can answer 'yes' to all three questions I have written this book primarily for you.

The main aim of this book is to encourage and develop your interest in statistics, which I have found to be a fascinating subject for over twenty years. Other aims are to help you to:

1. Understand the essential ideas and concepts of statistics.
2. Perform some of the most useful statistical methods.
3. Be able to judge which method is the most appropriate in a given situation.
4. Be aware of the assumptions and pitfalls of the methods.

Because of the wide variety of subject areas which require knowledge of introductory statistics, the worked examples of the various methods given in the main part of the text are not aimed at any one subject. In fact they deliberately relate to methods which can be applied to 'people data' so that every student can follow them without specialist knowledge. The end-of-chapter worksheets, on the other hand, relate to a wide variety of subjects to enable different students to see the relevance of the various methods to their areas of special interest.

You should tackle each worksheet before proceeding to the next chapter. To help with the necessary calculations you should be, or quickly become, familiar with an electronic hand calculator with the facilities given below.* (These facilities are now available on most scientific calculators.)

---

* *Calculators* The minimum requirements are: a memory, eight figures on the display, a good range of function keys (including square, square root, logarithm, exponential, powers, factorials) and internal programs for mean and standard deviation.

Answers and partial solutions are given to all the questions on the worksheets. When you have completed the whole book (except for the sections marked with an asterisk (*), which may be omitted at the first reading), a multiple-choice test is also provided, as a quick method of self-assessment.

TO THE TEACHER OR LECTURER

This book is not intended to do away with face-to-face teaching of statistics. Although my experience is that statistics is best taught in a one-to-one situation with teacher and student, this is clearly not practical in schools, colleges and polytechnics where introductory courses in statistics for non-specialist students often demand classes and lectures to large groups of students. Inevitably these lectures tend to be impersonal.

Because I have concentrated on the essential concepts and methods, the teacher who uses this book as a course text is free to emphasize what he or she considers to be the most important aspects of each topic, and also to add breadth or depth to meet the requirements of the particular course being taught.

Another advantage for the teacher is that, since partial solutions are provided to all the questions on the worksheets, students can attempt these questions with relatively little supervision.

WHAT THIS BOOK IS ABOUT

After introducing statistics as a science in Chapter 1 and statistical notation in Chapter 2, Chapters 3 and 4 deal with descriptive or summary statistics, while Chapters 5, 6 and 7 concentrate on probability and four of the most useful probability distributions.

The rest of the book comes broadly under the heading of statistical inference. After discussing sampling in Chapter 8, two branches of inference – confidence interval estimation and hypothesis testing – are introduced in Chapters 9 and 10 by reference to several 'parametric' cases. Three non-parametric hypothesis tests are discussed in Chapter 11.

In Chapters 12 and 13 association and correlation for bivariate data are covered. Simple linear regression is dealt with Chapter 14 and χ^2 goodness-of-fit tests in Chapter 15.

I have attempted throughout to cover the concepts, assumptions and pitfalls of the methods, and to present them clearly and logically with the minimum of mathematical theory.

Acknowledgements

The quotations given at the beginning of Chapters 1, 2, 3, 4, 8, 10, 11, and 14 are taken from a very interesting book on diseases and mortality in London in the 18th century. I would like to thank Gregg International, Amersham, England for permission to use these quotations from *An Arithmetical and Medical Analysis of the Diseases and Mortality of the Human Species*, by W. Black, 1973.

Acknowledgements for permission to use various statistical tables are given in Appendix C.

Thanks also to all the colleagues and students who have influenced me, and have therefore contributed indirectly to this book. Most of all I am grateful to my wife, Merilyn, for her support and encouragement throughout. I dedicate this new edition to her.

Chapter 1

What Is Statistics?

> Authors … have obscured the works in a cloud of figures and calculation: the reader must have no small portion of phlegm and resolution to follow them throughout with attention: they often tax the memory and patience with a numerical superfluity, even to a nuisance.

1.1 Statistics as a Science

You may feel that the title of this chapter should be 'What are statistics?', indicating the usual meaning of statistics as **numerical facts or numbers**. So, for example, the unemployment statistics that are published monthly might indicate the number of people registered as being unemployed during the month. However, in the title of this chapter, the singular noun **'Statistics'**, using the upper-case S, is used to mean the **science of collecting and analysing data**, where the plural noun 'data' means numerical or non-numerical facts or information.

We may collect data about 'individuals', that is individual people or objects. There may be many characteristics which vary from one individual to another. We call these characteristics **variables**. For example, individual people vary in height and unemployment status, and so height and unemployment status are variables.

Let us consider an example of some data which we might wish to analyse. Suppose our variable of interest is the height of first-year university students in the U.K. A Statistician might refer to these heights as a 'population' of heights.

Table 1.1 Data Set for a Random Sample of 40 Students

Student Reference Number	Sex 1 = Male 2 = Female	Height (cm)	Number of Siblings	Distance from Home to Oxford (km)	Type of Degree 1 = BA 2 = BSc	A-Level Count
1	1	183	1	80	2	6
2	2	163	2	3	1	32
3	2	152	2	90	1	22
4	2	157	3	272	2	12
5	2	157	1	80	2	12
6	2	165	3	8	2	18
7	1	173	1	485	2	14
8	1	180	2	176	2	8
9	2	164	2	10	2	6
10	2	160	3	72	1	18
11	2	166	0	294	2	16
12	2	157	1	22	1	12
13	2	168	0	144	2	12
14	2	167	2	160	2	12
15	2	156	1	50	2	10
16	2	155	1	64	1	12
17	1	178	1	224	2	8
18	2	169	3	480	2	10
19	2	171	5	56	2	6
20	1	175	3	141	2	8
21	1	169	2	259	2	4
22	2	168	4	96	2	6

23	2	165	1	104	2	12
24	2	166	1	90	2	8
25	2	164	3	72	2	22
26	2	163	1	37	2	8
27	2	161	1	208	2	6
28	2	157	2	40	2	18
29	1	181	2	120	2	10
30	2	163	1	400	2	10
31	2	157	2	208	1	10
32	2	169	2	169	2	12
33	2	177	2	410	1	16
34	2	174	1	90	2	10
35	1	183	1	80	2	10
36	1	181	2	278	2	8
37	1	182	1	240	2	6
38	1	171	9	192	2	24
39	1	184	2	35	1	14
40	1	179	1	45	1	10

We would expect these heights to vary. We could start by choosing one university from all those in the U.K., we could then choose 40 first-year students from the university's enrolment list, and we could measure the heights of these students (see Table 1.1). A Statistician might refer to these 40 heights as a sample (from the population of heights). There are many other ways of collecting and analysing such data. Indeed, this book is about how surveys like this should be conducted, and clearly they cannot be discussed in detail at this stage. It is, however, instructive to ask some of the questions which need to be considered before such a survey is carried out.

The most important question is 'What is the purpose of the survey?' The answer to this question will help us to answer other questions. How many students should be selected altogether? Is it better to choose all the students from one university or a number from each of a number of universities? How many should be selected from each of the chosen universities?

How should we select a given number of students from the enrollment list of a university? What do we do if a selected student refuses to cooperate in the survey? How do we allow for known or suspected differences between, for example, male and female student heights? Does the mean height of the students selected for the survey tell us all we need to know about their heights? How can we relate the mean height of the sample of selected students to the mean height of the population of heights, i.e., the heights of all first-year U.K. students?

The last question is an example of a general statistical method called *Statistical Inference*, which is one of the main branches of Statistics and also of this book.

1.2 Types of Statistical Data

Before we look at how data may be collected and analysed, we will consider the different types of statistical data we may need to study. As stated in the Preface to the First Edition, the main part of this book will be concerned with 'people data', for example, the data in Table 1.1, which gives information about six variables for a sample of 40 students, namely:

Sex
Height
Number of siblings (brothers and sisters)
Distance from home to Oxford
Type of degree
A-level count

Some of these variables are **categorical**, that is, the 'value' taken by the variable is a nonnumerical category or class. An example of a categorical variable is sex, with two categories, male and female. Some variables are quantifiable, that is, they may take numerical values. These numerical variables can further be classified as being continuous, discrete, or ranked using the following definitions:

A **continuous** variable can take any value in a given range.
A **discrete** variable can take only certain distinct values in a given range.
A **ranked** variable is a categorical variable for which the categories imply some order or relative position.

Example

Height is an example of a continuous variable since an individual adult human being may have a height anywhere in the range 100 to 200 cm. We can usually decide that a variable is continuous if it is **measured** in some units.

Example

Number of brothers and sisters (siblings) is an example of a discrete variable, since an individual human can have 0, 1, 2, ... siblings, but cannot have 1.43, for example. We can usually decide that a variable is discrete if it can be **counted**.

Example

Birth order is an example of a ranked variable, since an individual human may be the first-born, second-born, etc., into a family, with a corresponding birth order of 1, 2, etc.

Table 1.2 below shows the results of applying similar ideas to all the variables in Table 1.1.

Table 1.2   Examples of Types of Statistical Data

Name of Variable	Type of Variable	Likely Range of Values or List of Categories
Sex	Categorical	Male, female
Height	Continuous	100 to 200 cm
Number of siblings	Discrete	0, 1,...,10
Distance home to Oxford	Continuous	1 to 500 km
Type of degree	Categorical	BA, Bsc
A-level count	Discrete	0, 1, 2,...,50

The distinction between the continuous and the discrete variable is, in practice, not as clear-cut as stated above. For example, most people give their age as a whole number of years, so that age appears to be a discrete variable which increases by one at each birthday. The practice of giving one's age approximately, for whatever reason, does not alter the fact that age is fundamentally a continuous variable.

Now try Worksheet 1.

Worksheet 1: Statistic(s) and Types of Statistical Data

1. Which of the following is a continuous variable?
   (a) In a driving test, the time between the examiner saying 'STOP' and the car coming to a halt.
   (b) The colour of the car in which a learner-driver takes the test.
   (c) The number of times a learner-driver takes the driving test before passing.
2. For the following 17 cases, decide whether the variable is continuous, discrete, ranked, or categorical. Give a range of likely values or a list of categories.

   The value or category of the variable varies from one 'individual' to another. The individual may or may not be human, as in question (h) below, where the individual is 'county'. Name the individual in each of the 17 cases.
   (a) The number of current account balances checked by a firm of auditors each year.
   (b) The present cost of bed-and-breakfast in 3-star London hotels.
   (c) The occupation type of adult males.
   (d) The number of failures per 100 hours of operation of a large computer system.
   (e) The number of hours lost per 100 hours due to failure of a large computer system.
   (f) The number of cars made each month by a car manufacturer.
   (g) The position of the British entry in the annual Eurovision song contest.
   (h) The annual rainfall in English counties in 1993.
   (i) The number of earthquakes per year in a European country in the period 1900–1999.
   (j) The number of times rats turn right in 10 encounters with a T-junction in a maze.
   (k) The grades obtained by candidates taking A-level mathematics.
   (l) The colour of a person's hair.

(m) The presence or absence of a plant species in each square metre of a meadow.

(n) The reaction time of rats to a stimulus.

(o) The yield of tomatoes per plant in a greenhouse.

(p) The constituents found in core samples when drilling for oil.

(q) The political party people vote for in an election.

3. Consider again the variable 'age', which we have seen could be continuous or discrete. Show how age could also be a categorical variable, and state the categories. Can these categories be put into a logical ranking order?

4. (a) Must a ranked variable be categorical? Explain.

   (b) Is there a logical ranking order to either of the following categorical variables: 'sex' and 'type of degree'?

   (c) Think of another categorical variable for which the categories must clearly (i) be ranked, (ii) not be ranked.

5. (a) Using any research method available to you (the Internet ?), find the earliest reference to the words

   (i) STATISTIC or STATISTICS.

   (ii) STATIST.

   (b) What is or was a STATIST?

Chapter 2

Some Statistical Notation

I have corrected several errors of preceding calculators....

2.1 Introduction

It is not necessary for you to master all the notation in this chapter before proceeding to Chapter 3. However, references to this notation will be made in later chapters within the context of particular statistical methods. Worksheet 2, which follows this chapter, is intended to help you to use your calculator and become familiar with the notation. Nowadays, some statisticians use only computers to help solve statistical problems. However, learning first by calculator is, in my opinion, preferable because the 'student' is then more likely to understand the underlying calculations and concepts.

2.2 Σ

The symbol Σ (the upper-case version of the Greek letter sigma) implies the operation of summation. If x stands for a variable, then Σx means 'sum all the observed values of x'. If there are n observations in a sample taken from a population, then we can write:

$$\text{Sample mean of } x = \frac{\text{sum of the observed values of } x}{\text{number of observed values}}$$

This can be written in symbols as:

$$\bar{x} = \frac{\Sigma x}{n} \tag{2.1}$$

We pronounce $\bar{x}$ as 'x bar'. You will find $\bar{x}$ on any scientific calculator, while Minitab simply uses the word 'mean'.

Example

The sample of five coins in my pocket have the following values (p):

$$1, 2, 2, 5, 100.$$

So we can write:

$$\text{Sample mean} = \frac{1 + 2 + 2 + 5 + 100}{5} = 22p$$

Or, using Formula (2.1), $n = 5$, $\Sigma x = 110$, $\bar{x} = 110/5 = 22p$
Other uses of the Σ notation are Σx^2, $(\Sigma x)^2$, and $\Sigma(x - \bar{x})$, defined as follows:

(Σx^2) means square n observations of x and then sum.   (2.2)

$(\Sigma x)^2$ means sum the n observed values of x and then square this sum   (2.3)

$\Sigma(x - \bar{x})$ means subtract the sample mean from each observed value of x and then sum.   (2.4)

Example

Carry out the above operations on the data in the previous example:

$$\Sigma x^2 = 1^2 + 2^2 + 2^2 + 5^2 + 100^2 = 10,034, \quad \textit{units are } p^2.$$
$$(\Sigma x)^2 = (1 + 2 + 2 + 5 + 100)^2 = 12,100$$
$$\Sigma(x - \bar{x}) = (1 - 22) + (2 - 22) + (2 - 22)$$
$$+ (5 - 22) + (100 - 22) = 0.$$

Note that $\Sigma(x - \bar{x})$ will always be zero for any set of sample data.

2.3   Factorials

If n is a positive integer (whole number), then $1 \times 2 \times 3 \times \ldots \times n$ is called **factorial** n and is written $n!$ So we can write:

$$n! = 1 \times 2 \times 3 \times \ldots\ldots \times n$$

(n must be a positive integer)

(2.5)

Examples

$3! = 1 \times 2 \times 3 = 6$
$5! = 1 \times 2 \times 3 \times 4 \times 5 = 120$
$1! = 1$

Try these examples on your calculator.

In addition to the above definition of factorial n, factorial 0 is defined as 1, so $0! = 1$. Try this on your calculator as well.

Remember that factorials for any other numbers are not defined. So $-5!$ and $2.3!$ are not defined and hence are meaningless.

Applications of the above factorial notation will initially be used in this book in the calculation of 'binomial probabilities' in Chapter 6.

2.4 x^y

To find the 'power y of any number x' you need the x button on your calculator.

Examples

$(0.6)^4$ implies $x = 6$, $y = 4$. $(0.6)^4 = 0.6 \times 0.6 \times 0.6 \times 0.6 = 0.1296$.
  Check this on your calculator using the x^y button.
$(0.6)^0$ implies $x = 0.6$, $y = 0$. So $(0.6)^0 = 1$.
  Check this on your calculator.
  The x^y button is useful in Chapter 6 in calculating binomial probabilties.

2.5 e^x

The letter e in mathematics and on your calculator stands for the number 2.718
approximately. We need to be able to obtain values of e^x in Chapter 6 in the
calculation of Poisson probabilities.

Examples

$e^1 \; = e = 2.718$

$e^{-2} = 0.1353$

$e^0 \; = 1$

Try these on your calculator.

2.6 Decimal Places and Significant Figures

Calculators produce many figures on the display and it is tempting to
write them all down. You will learn by experience how many figures are
meaningful in an answer. For the moment, concentrate on giving answers
to a stated number of decimal places or significant figures.

  Use the idea that, for example, 3 decimal places (dps) means write
three figures only to the right of the decimal point, rounding the third
figure (after the decimal point) up if the fourth figure is 5 or more.

Examples

  1.6666 to 3 dps is 1.667
  1.6665 to 3 dps is 1.667
  1.6663 to 3 dps is 1.666
  1.677 to 3 dps is 1.670
  167 to 3 dps is 167.000

The number of significant figures (sfs) means the number of figures (as you scan from left to right) starting with the first non-zero figure. Round the last significant figure up if the figure immediately to its right is 5 or more. Nonsignificant figures to the left of the decimal point are written as zeros, while those to the right of the decimal point are omitted.

Examples

26243   to 3 sfs is 26200

2624   to 3 sfs is 2620

2626   to 3 sfs is 2630

26.24   to 3 sfs is 26.2

0.2624   to 3 sfs is 0.262

0.002626 to 3 sfs is 0.00263

Worksheet 2: Some Statistical Notation

1. Check that you are able to work out each of these on your calculator:
    (a) $1.3 + 2.6 - 5.7$
    (b) $10.0 - 3.4 - 2.6 - 1.0$
    (c) $(2.3)(14.6)$
    (d) $(0.009)(0.0274)(1.36)$
    (e) $2.3/14.6$
    (f) $1/0.00293$
    (g) $(2.3 + 4.6 + 9.2 + 17.3)/4$
    (h) $28^{0.5}$
    (i) $(0.5)^3$
    (j) $(0.2)^2 (0.8)^4$
    (k) $(0.5)^0$
    (l) $(0.2)^{-3}$
    (m) $e^{1.6}$
    (n) $e^{-1.6}$
    (o) $\dfrac{13}{\sqrt{10 \times 24}}$
    (p) $6 - (-0.5)(4)$
    (q) $4!, 1!, 6!, (-3)!, (2.4)!$
2. Express your answer to Question:
    (a) 1(c) to 1 dp
    (b) 1(d) to 2 sfs
    (c) 1(e) to 2 sfs

    (d) 1(f) to 4 sfs

    (e) 1(f) to 1 sf

3. Use the memory facility on your calculator to work out the following:

    (a) $1 + 2 + 3 + 4 + 5 + 6 + 7 + 8 + 9 + 10$

    (b) $(1 + 2 + 3 + 4 + 5)/5$

    (c) $1^2 + 2^2 + 3^2 + 4^2 + 5^2$

    (d) $(1 \times 2) + (3 \times 4) + (5 \times 6)$

4. For the eight observed values of x: 2, 3, 5, 1, 4, 3, 2, 4, find

$$\Sigma x, \ \bar{x}, \ (\Sigma x)^2, \ \Sigma(x - \bar{x}), \ \Sigma(x - \bar{x})^2, \quad \text{and} \quad \Sigma x^2 - \frac{(\Sigma x)^2}{n}$$

5. Repeat Question 4 for the five observed values of x: 2.3, 4.6, 1.3, 7.2, and 2.3.

6. In Questions 4 and 5, you should find that $\Sigma(x - \bar{x}) = 0$ in both cases, and that $\Sigma(x - \bar{x})^2 = \Sigma x^2 - \frac{(\Sigma x)^2}{n}$, also in both cases.

    Verify that these two results hold for the data in Questions 4 and 5 above.

    Can you prove that the two results hold for any set of sample data?

Chapter 3

Summarizing Data by Tables and by Graphical Methods

The important data …. are condensed, classed, and arranged into concise tables.

3.1 Introduction

If we collect data, it is often a good idea to use tabular and graphical methods to 'explore' the data before we do any calculations. Several examples will be given using all the types of data discussed in Chapter 2. Initially we will concentrate on one-variable data, but later bivariate (two-variable) data will be considered.

3.2 Tables and Graphs for One Continuous Variable

The third column of Table 1.1 gives the heights of a sample of 40 students. These heights may be rewritten as shown in Table 3.1. We may represent these data graphically in several different ways, for example:

    (a)   a **dotplot** (Fig. 3.1),
    (b)   a **stem and leaf display** (Fig. 3.2),
    (c)   a **box and whisker plot** (Fig. 3.3).

Table 3.1 List of the Heights (cm) of 40 Students

183	163	152	157	157	165	173	180	164	160
166	157	168	167	156	155	178	169	171	175
169	168	165	166	164	163	161	157	181	163
157	169	177	174	183	181	182	171	184	179

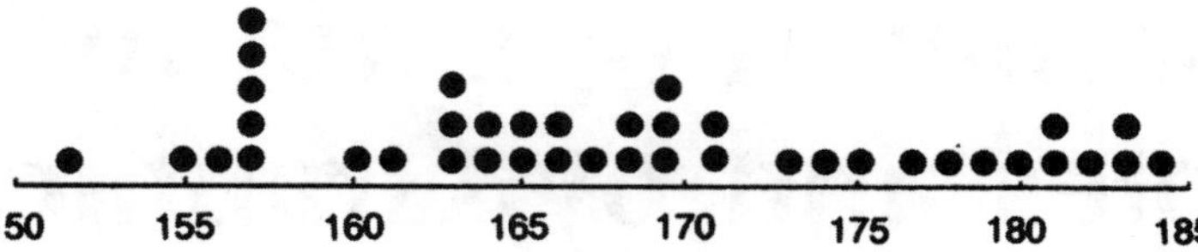

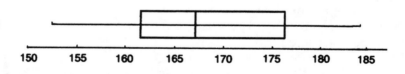

Figure 3.1 Dotplot for the Data in Table 3.1

```
15 | 2
15 | 5 6 7 7 7 7 7
16 | 0 1 3 3 3 4 4
16 | 5 5 6 6 7 8 8 9 9 9
17 | 1 1 3 4
17 | 5 7 8 9
18 | 0 1 1 2 3 3 4
```

Figure 3.2 Stem and Leaf Display for the Data in Table 3.1

Figure 3.3 Box and Whisker Plot for the Data in Table 3.1

The interpretation of Fig. 3.1 is relatively straightforward. Each observation is represented by one dot on the scale of the variable, which is height in this case. Looking at the dotplot, we see that the dots are fairly evenly spread across the range 155 to 185 cm, with perhaps a tendency to bunch more in the range 163 to 170 cm. The dots are more or less symmetrically distributed about a 'middle' value of approximately 167 cm.

The stem and leaf display is a way of representing the data in what is a mixture of a graph and a table. In Fig. 3.2, the column of numbers to the left of the vertical line is the 'stem', while values to the right of the line are the 'leaves'. The first row in Fig. 3.2 is for observations from 150 to 154 inclusive, while observations from 155 to 159 go in the second row, and so on. Note that the leaves are written in rank order. You need to turn Fig. 3.2 through 90 degrees to compare its shape with Fig. 3.1. The interpretation is similar to that above for the dotplot.

The box and whisker plot is the hardest to interpret at this stage because we do not know what the box (i.e., the rectangle) represents, and the same goes for the whiskers (i.e., the horizontal line through the box). In fact, the vertical line which divides the box into two corresponds to the median value for the variable (167.5 for our data), the ends of the whiskers correspond to the minimum (smallest) and maximum (largest) values (152 and 184, respectively, for our data), while the points where the whiskers intersect with the box correspond to the lower and upper quartiles (161.5 and 176.5, respectively, for our data). We will meet and define the terms: median, lower quartile, and upper quartile in Chapter 4.

The 40 observations in Table 3.1 can be grouped as shown in Table 3.2, which is an example of a **grouped frequency distribution table**. The groups, 149.5 to 154.5 and so on, have been decided using the following guidelines:

(a) There are between 5 and 10 groups for smallish data sets (and up to 15 groups for large data sets, e.g., where the total frequency is above 500 and where frequency means 'number of observed values'). If there are too few groups, it is difficult to see how the

Table 3.2   Grouped Frequency Distribution for the Heights (cm) of 40 Students

Height	Number of Students (Frequency)
149.5 to 154.5	1
154.5 to 159.5	7
159.5 to 164.5	7
164.5 to 169.5	10
169.5 to 174.5	4
174.5 to 179.5	4
179.5 to 184.5	7
Total	40

data vary (i.e., the 'distribution of the data'). If there are too many groups, then the table is less of a summary.

(b) Each observation must go into one and only one of the groups. For example, it is clear that the number 160 would go into the third group, while 159 would go into the second group.

(c) The groups are equally wide, unless there is a very good reason why they should be unequal. In Table 3.2, each group is 5 cm wide, for example (154.5 − 149.5) = 5. It is also easier to represent data graphically if the groups are equally wide. Table 3.2 can be represented graphically in the form of a **histogram** (Fig. 3.4), noting that the midpoint of the first group is (149.5 + 154.5)/2 = 152, and so on.

Note that the vertical axis of the histogram represents frequency only if the groups are of equal width, as they are in this example.

The interpretation of Fig. 3.4 has to be the same as for Fig. 3.2 since they have identical shapes! Try turning Fig. 3.2 through 90 degrees counterclockwise, and you will see this for yourself.

The next table in this section is the **cumulative frequency distribution** table (Table 3.3), which we derive from Table 3.2. The values in the height column are group endpoints (it is a common mistake to use midpoints instead). The table provides information such as '8 students have a height of less than 159.5 cm'.

Table 3.3 can be represented graphically in the form of a **cumulative frequency polygon**. Notice that each row of Table 3.3 gives rise to a point on Fig. 3.5, starting with a cumulative frequency of zero and ending with a cumulative frequency equal to the total frequency (40 in the example).

It is also a common mistake for students to use group midpoints rather than group endpoints in drawing a cumulative frequency polygon.

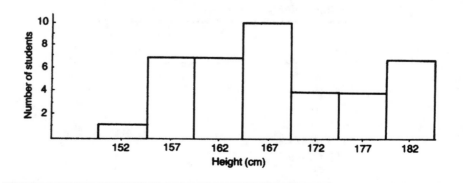

Figure 3.4   Histogram for the Data in Table 3.2

Table 3.3 Cumulative Frequency Distribution Table for the Heights (cm) of 40 Students

Height	Cumulative Number of Students (Cumulative Frequency)
149.5	0
154.5	1
159.5	8 (= 1 + 7)
164.5	15
169.5	25
174.5	29
179.5	33
184.5	40

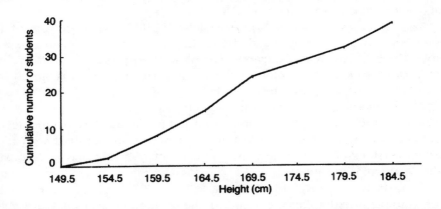

Figure 3.5 Cumulative Frequency Polygon for the Data in Table 3.3

3.3 Using Minitab for Windows to Draw Graphs

In this section we will see how to use Minitab to draw four of the graphs introduced earlier in this chapter, namely the dotplot, the stem and leaf display, the box and whisker plot, and the histogram. We will be using the 'height' data shown in Table 1.1, but this time we will assume that the data from the six columns in Table 1.1 which hold the data on sex, height, number of siblings, distance from home to Oxford, type of degree, and A-level count have been stored in a Minitab file (called ES4DATA.MTW), which can be retrieved as necessary. Let's assume this file is stored on a floppy disc, so that it can be loaded into the A-drive of your computer.

Notes

(a) If you are not familiar with Minitab for Windows data entry, data storage and retrieval, you should read Appendix E before proceeding.

(b) If you are not familiar with any Windows applications at all, you should consult an appropriate book, such as the 540 page! book called *The Student Edition of Minitab for Windows*, by John McKenzie, Addison Wesley Longman (1995), ISBN 0-201-59886-8. The version of Minitab used by McKenzie et al. is the version used throughout this (the fourth) edition of *Essential Statistics*.

Now proceed as follows:

Choose **File** > **Open Worksheet** > **Minitab Worksheet** > **Select File**.

Then, changing to the A-drive if necessary, enter **ES4DAT.MTW** in the box below **File Name**. Click on **OK**.

You should soon see a Window called DATA, which consists of a spreadsheet with columns headed C1, C2,…, and rows labelled 1, 2,…

Now: Choose **Graph** > **Character Graphs** > **Dotplot**.

In order to draw a dotplot for the variable Height, which is stored in C2: enter **C2** in the **Variable** box. Click on **OK**.

A dotplot should appear on the screen.

Repeat the various steps above substituting in turn: stem and leaf, boxplot, histogram, instead of dotplot. When you are satisfied with the screen versions of the four graphs, you can obtain a printout by: Choose **File** > **Print Window**.

Compare your four graphs with Fig. 3.1, Fig. 3.2, Fig. 3.3, and Fig. 3.4. For example, the dotplots are identical; the stem and leaf displays are the same except for the extra column to the left in Minitab's display — these are frequencies and will be referred to again in Chapter 4.

The boxplots are the same except that there are no numbers on the height scale in Minitab's boxplot. This can be put right by entering 150 and 190 as the minimum and maximum positions in the relevant boxes in the boxplot window.

The two histograms look very different, although they are telling the same story. For a histogram like Fig. 3.4, use the following:

Choose **Graph** > **Histogram**.

Enter **C2** in col 1, row 1.

Click on **Options** (to reveal histogram options window).

Enter **152:182/5** in the box called '**Define intervals using values**' (152 and 182 are the midpoints of the first and last groups in Fig. 3.4, and 5 is the width of each group in the same figure).

Click on **OK**.

Note

You will probably have noticed that, under the heading Graph in Minitab, there are two lists of graphs, the first list contains the names of High Resolution (or 'Professional') graphs. The second list contains the names of Character graphs.

3.4 Tables and Graphs for One Discrete Variable

In Table 1.1, 'number of siblings' is an example of a discrete variable. Table 3.4 shows these raw data, from which a dotplot would be a reasonable graph to draw (see Fig. 3.6).

None of the other three graphs used in the previous section for the continuous variable height are useful for the discrete variable 'number of siblings'. Why is this? For the stem and leaf display, there are not enough different values for the number of siblings, only 10, i.e., 0, 1,…, 9, while for height there were about 40 different values.

For the boxplot, we need to discuss the median, etc. (which we will in the next chapter), while the histogram should, in my view, be used only for continuous variables because it is a continuous picture.

A graph for 'sibs' should have gaps, and once a frequency distribution table has been drawn up (see Table 3.5) the Line Chart, e.g., Fig. 3.7, follows naturally. The interpretation of Fig. 3.6, and Fig. 3.7 is that the most popular number of siblings is 1, closely followed by 2. I would think that the average number of siblings is between 1 and 2. Averages will be discussed in Chapter 4. There are very few cases in which there are either no siblings ('the only child') or more than three siblings.

Table 3.4   The Number of Siblings for 40 Students

1	2	2	3	1	3	1	2	2	3
0	1	0	2	1	1	1	3	5	3
2	4	1	1	3	1	1	2	2	1
2	2	2	1	1	2	1	9	2	1

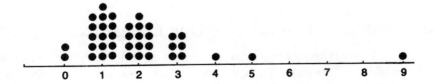

Figure 3.6   Dotplot for the Data in Table 3.4

Table 3.5 Grouped Frequency Distribution for the Number of Siblings of 40 Students

Number of Siblings	Number of Students (Frequency)
0	2
1	16
2	13
3	6
4	1
5	1
6	0
7	0
8	0
9	1

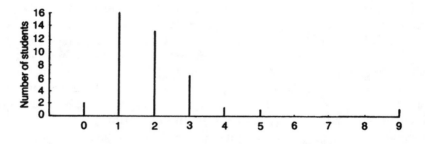

Figure 3.7 Line Chart for the Number of Siblings of 40 Students

Minitab does not distinguish between continuous and discrete data! For the 'sibs' data, a Minitab dotplot may be drawn (as in Section 3.3), but Minitab does not do a Line Chart. The nearest Minitab graph is obtained by using: choose **Graphs > Character Graphs > Histogram**.

The rest is left as an exercise for the reader!

3.5 Tables and Graphs for One Categorical Variable

In Table 1.1, Sex is an example of a categorical variable, with two categories, male and female. These have been converted to 1 and 2, respectively, partly for convenience, but mainly because Minitab will only accept numerical data. The raw data for Sex are shown in Table 3.6, grouped data are shown in Table 3.7, which gives rise to a Bar Chart, Fig. 3.8.

Table 3.6   The Sex of 40 Students (1 = Male, 2 = Female)

1	2	2	2	2	2	1	1	2	2
2	2	2	2	2	2	1	2	2	1
1	2	2	2	2	2	2	2	1	2
2	2	2	2	1	1	1	1	1	1

Table 3.7   A Grouped Frequency Table for the Sex of 40 Students

Sex	Number of Students (Frequency)
Male	13
Female	27

Figure 3.8   Bar Chart for the Data in Table 3.7

The interpretation of Fig. 3.8 and of Table 3.7 is that about 1/3 of students are male and 2/3 are female (or you could say that there are twice as many females as there are males).

Since Minitab sees this type of input as a special case of grouped data with only two possible values, the only Mintab graphs (of the four drawn for the variable 'height') is the dotplot. (Although you could try: Choose **Graphs > Character Graphs > Histogram,** etc. as in Section 3.4).

3.6   Tables and Graphs for Two-Variable Data

These types of data wll not be discussed in detail in this section, but a few specific examples will be given because they will be important in later chapters.

When both variables are categorical, the frequencies with which the various cross-categories occur can be displayed in a two-way table, often referred to as a **contingency table**. Table 3.8 is an example of a 2 × 2 contingency table. The categories of one of the variables, i.e., Sex, are the row names, namely, male and female, while the categories of the other variable, i.e., type of degree, are the column names, namely, BA and BSc. The numbers in the four 'cells' are the frequencies of the four cross-categories.

These are easily obtained by referring to the relevant columns of Table 1.1. The question of interest for these types of data is usually, 'Is there some association between the variables (e.g., Sex and Type of degree), or are the variables independent of one another? Questions like this will be discussed in Chapter 13. What conclusions would you draw, assuming that you haven't read Chapter 13? This is left as an exercise for the reader.

Table 3.8   Sex and Type of Degree for 40 Students

	Type of degree	
Sex	BA	BSc
Male	2	11
Female	7	20

Minitab can produce Table 3.8 as follows:
Choose **Stat** > **Tables** > **Cross Tabulation**.
Enter **C1** and **C5** in the **Classification Variables** box.
Click on the box to the left of **Counts**.
Click on **OK**.

When both variables are continuous, the raw data may be held as two columns, for example, as height and distance are in Table 1.1. A very useful graphical method in this case is the scatter diagram. Fig. 3.9 shows a **scatter diagram** of distance (Y axis) against height (X axis).

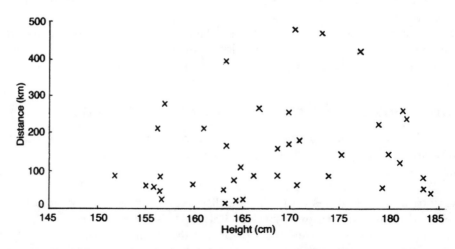

Figure 3.9   Scatter Diagram of the Heights and Distances from Home for 40 Students

This diagram shows very little pattern, in fact it looks fairly random. In Chapters 14 and 15 we will be interested in cases where the points appear to exhibit a linear trend, i.e., tend to lie on a straight line, albeit with some scatter of points about the line. We will be asking questions such as 'Is there a significant association or correlation between the two variables?' If the answer is 'Yes', how well can we predict one variable from the other? As an example think about a scatter diagram where one variable is maximum temperature on a summer's day at a holiday resort, while the other variable is the daily amount of ice-cream bought each day at the resort. In this case you might expect that, as the temperature increased, so would the sales of ice-cream, and vice versa. The Minitab method for obtaining a scatter diagram (which is referred to as a scatter *plot* by Minitab) is as follows:

Choose **Graph** > **Character Graphs** > **Scatter Plot**.
Enter **C4** in the **Y variable** box.
Enter **C2** in the **X variable** box.
Click on **OK**.

When one variable is continuous and the other is discrete (with only a few possible values) or categorical (with only a few categories), it is often a good idea to plot a number of dotplots of the continuous variable for each value or category of the other variable. For example, we may wish to compare graphically the heights of male and female students (see Fig. 3.10).

Notice that these dotplots have the same scale for height. This makes it much easier to compare them, but the interpretation of Fig. 3.10 is not so easy. An 'eye-ball' inspection gives the impression that male heights, on average, are greater than female heights, although there is some overlap between the two data sets. The question of interest is 'Is the apparent difference in the two sets of height data a chance difference or a real difference?' Questions like this will be discussed in depth from Chapter 9 onwards.

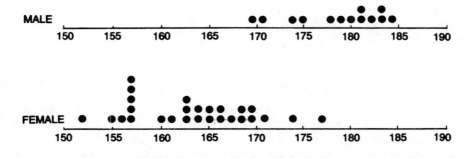

Figure 3.10   Dotplots of the Heights of Male and Female Students

The following shows how to get Minitab to produce two (or more) dotplots on the same scale:

Choose **Graph** > **Character Graphs** > **Dotplot**.
Enter **C2** in the **Variables** box.
Click in the box to the left of the words **'By variable'**.
Enter **C1** in the box to the right of **'By variable'**.
Click in the box to the left of the words **'Same scale for all variables'**.
Click on **OK**.

3.7 Summary

When one-variable or two-variable data are collected for a number of individuals or subjects, these data may be summarised in tables or graphically. Some form of grouping may be advisable if there are many observations; the particular type of table and graph used to summarise the data depends on the type(s) of variable(s). Examples discussed in this chapter are shown in Table 3.9.

Worksheet 3: Summarizing Data by Tables and by Graphical Methods

1. Decide which type of table and graphical method you would use on the following one-variable data sets:
    (a) The type of occupation of 50 adult males.
    (b) The total number of earthquakes recorded in the 20th century for each of 10 European countries.
    (c) The percentage of ammonia converted to nitric acid in each of 50 repetitions of an experiment.
    (d) The number of hours of operation in a given month for 49 nominally identical computers.
    (e) The number of right turns made by 100 rats, each rat having 10 encounters with T-junctions in a maze.
    (f) The systolic blood pressure of 80 expectant mothers.
    (g) The number of errors (assume a maximum of 5) found by a firm of auditors in 100 balance sheets.
    (h) The number of each of six types of room in a large hotel. The types are single-bedded, double-bedded, single and double bedded, each with or without bath.
    (i) The density of 10 blocks of carbon dioxide.
    (j) The number of sheep farms on each type of land. The land types are flat, hilly, and mountainous.

Table 3.9   Types of Table and Graph Used to Summarize Data

Number of Variables	Variable Type	Type of Table (and Reference)	Type of Graph (and Reference)
One	Continuous	Ungrouped (Table 3.1)	Dotplot (Fig. 3.1) Stem and leaf display (Fig. 3.2) Box and whisker plot (Fig. 3.3)
		Grouped frequency (Table 3.2)	Histogram (Fig. 3.4)
		Cumulative frequency (Table 3.3)	Cumulative frequency polygon (Fig. 3.5)
	Discrete	Ungrouped (Table 3.4)	Dotplot (Fig. 3.6)
		Grouped frequency (Table 3.5)	Line chart (Fig. 3.7)
	Categorical	Ungrouped (Table 3.6)	
		Group frequency (Table 3.7)	Bar chart (Fig. 3.8)
Two	Both categorical	Contingency table (Table 3.8)	
	Both continuous	Two columns of Table 1.1	Scatter diagram (Fig. 3.9)
	One continuous, one categorical or discrete	Two columns of Table 1.1	Dotplots (Fig. 3.10)

(k) The fluoride content of the public water supply for 100 cities in the U.K.

2. The amounts of coffee in grams by which 70 jars of coffee exceeded the nominal 200 g were as follows:

0.7	1.3	1.4	2.2	1.6	0.8	1.2	3.2	2.3	4.6
1.9	1.7	0.2	2.0	2.3	3.1	0.6	2.7	2.9	2.8
1.1	0.7	2.8	1.3	0.3	1.6	3.3	0.4	0.6	5.7
2.3	1.3	2.1	0.9	1.5	2.1	0.9	1.8	3.5	3.5
0.5	2.8	1.6	2.2	0.9	1.2	3.7	1.8	2.0	4.0
1.4	2.7	1.6	2.2	1.1	1.7	1.3	3.4	1.7	3.1
3.0	1.6	0.7	1.8	2.9	1.7	2.2	1.3	2.5	2.7

Draw a dotplot, and comment on the resulting distribution. Summarise the data in a grouped frequency table and draw a histogram. Comment on its shape. Using Minitab for Windows obtain a dotplot, a stem and leaf display, and a histogram. Which of these three graphs do you like the most for these data?

3. For the 'Distance data' in Column 5 of Table 1.1, draw a histogram using groups 0 to 49.9, 50 to 99.9, and so on. Now draw up a cumulative frequency table and polygon. If half the students live less than X km from Oxford, what is the value of X? Compare your answer with that obtained from the 'raw data' in Table 1.1.

4. Present graphs to help you to answer the question: 'Is the A-level count of Science students more or less the same as the A-level count of Arts students?'

5. The number of goals scored by each team in the 43 games of soccer played on 28/9/98 in the Carling Premiership league or one of the three Nationwide first, second, or third divisions were as follows:

Carling Premiership	Nationwide		
	Division 1	Division 2	Division 3
1 - 0	3 - 0	2 - 0	4 - 2
1 - 1	2 - 1	1 - 1	3 - 0
2 - 0	5 - 2	2 - 1	1 - 3
0 - 0	2 - 2	2 - 1	1 - 1
2 - 0	2 - 0	1 - 2	2 - 2
1 - 0	4 - 1	0 - 1	1 - 1
3 - 3	1 - 1	2 - 1	1 - 2
1 - 1	2 - 2	2 - 2	2 - 0
	1 - 0	3 - 1	0 - 0
	1 - 0	1 - 1	2 - 1
	1 - 0	1 - 3	1 - 1
	—	1 - 0	3 - 1

(a) Considering only the number of goals scored by each team, form a frequency distribution table for each of the 4 leagues, and also one for all 86 teams, irrespective of league.

(b) Draw five suitable graphs to represent these data, and comment on them.

(c) What is the average number of goals scored (i) per team, (ii) per match? Do this separately for each league, and comment.

(d) It is well known that the home team has an advantage, other things being equal. Remembering that, for example, a score of 3 to 2 means that the home team scored 3 goals, while the away team scored only 2 goals, what is the apparent average home advantage in each of the 43 games? Form one grouped frequency table for 'home advantage' for all 43 matches, and calculate the average home advantage.

Chapter 4

Summarizing Data by Numerical Measures

Let us condense our calculations into a few general abstracts......

4.1 Introduction

You are probably familiar with the word 'average' and you may have heard the term 'standard deviation'. Average and standard deviation are examples of numerical measures we use to summarise data. There are many other such measures. It is the purpose of this chapter to show how we may obtain some of these measures from a given data set, but it is equally important for you to learn when to use a particular measure in a given situation.

4.2 Averages

In this book, the word **average*** will be thought of as a vague word meaning 'a middle value' or better 'a single value which in some way represents all the data.' It will only take on a definite meaning if we

---

* The word 'average' was used in Section 2.2 and also in Question 5(c) of Worksheet 3. My guess would be that you would have taken it to mean the 'sample mean' as described in Section 4.3.

decide that we are referring to a rigorously defined measure such as the

(a) Sample (arithmetic) mean, or

(b) Sample median, or

(c) Sample mode.

Averages will be discussed in Sections 4.2 to 4.6 inclusive.

4.3 Sample Mean ($\bar{x}$)

The sample arithmetic mean, which we will refer to simply as the **sample mean** of a variable x, is defined in words as follows:

$$\text{sample mean of } x = \frac{\text{sum of the observed values of } x}{\text{number of observed values}}$$

The symbol we use for the sample mean is $\bar{x}$, and its definition in symbols is as follows:

$$\bar{x} = \frac{\Sigma x}{n} \tag{2.1}$$

where Σx means the 'sum of the observed values of x,' and n is the 'number of observed values.'

This formula first occurred in Section 2.2.

Example

The heights of a sample of 40 students are listed in Table 1.1. The sample mean height is

$$\bar{x} = \frac{183 + 163 + \ldots\ldots\ldots + 184 + 179}{40}$$

$$= \frac{6730}{40}$$

$$= 168.3 \, \text{cm}$$

The sample mean height of the 40 students is 168.3 cm. Note that we have used one more significant figure than for the raw data (in Table 1.1).

Formula (2.1) can be used for both continuous and discrete data, but not for categorical data since the term 'sample mean sex,' for example, has no meaning.

Assuming the data in Table 1.1 are stored in a Minitab file called ES4DAT.MTW, as it was in Section 3.3, you can go into Minitab, and:

Choose **File > Open Worksheet**

Enter **ES4DAT.MTW**

Choose **Stat > Basic Stats > Descriptive Stats**

Enter **C2 (or Height)** in the **Variables** box
Click on **OK**.

Minitab's output includes the following, which we will call Table 4.1:

Table 4.1 Summary Statistics for the Heights (in centimeters) of 40 Students Using Minitab

N	Mean	Median	Stdev	Min	Max	Q1	Q3
40	168.25	167.50	9.11	152.00	184.00	161.50	176.5

Minitab's value for the mean agrees with the value of 168.3 obtained 'by hand' to 1 dp. We will come across the other information in Table 4.1 later in this chapter.

4.4 Sample Median

The **sample median** of a variable x is defined as the middle value when the n sample observations of x are **ranked** in increasing order of magnitude.

$$\text{Sample median is the } (n + 1)/2^{\text{th}} \text{ value} \qquad (4.1)$$

Example: n odd

The heights of five students are 183, 163, 152, 157, and 157 cm.
   In rank order: 152, 157, 157, 163, and 183.
   Here $n = 5$, $(n + 1)/2 = 3$, so the median height is the third value, and is equal to 157 cm.

Example: n even

The heights of four students are 165, 173, 180, and 164 cm.
   In rank order: 164, 165, 173, and 180.
   Here $n = 4$, $(n + 1)/2 = 2.5$, so the median height is the mean of the second and third values, i.e., $(165 + 173)/2 = 169$ cm.

Example

Find the median height of the 40 student heights in Table 1.1. Instead of ranking these 40 values, we can more elegantly use one of the following four methods noting that, since $n = 40$ and hence $(n + 1)/2 = 20.5$, we

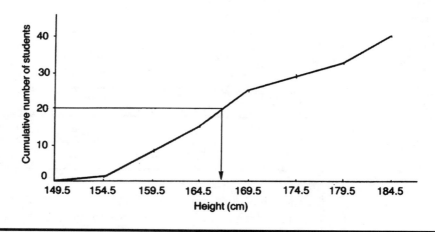

Figure 4.1   Cumulative Frequency Polygon for the Data in Table 3.3

want the mean of the 20th and 21st values, assuming the data are in rank order:

Method (a) draw a dotplot (as in Fig. 3.1);
Method (b) draw a stem and leaf display (as in Fig. 3.2);
Method (c) draw a cumulative frequency polygon (Fig. 3.5);
Method (d) use Minitab.

Both Fig. 3.1 and Fig. 3.2 show that the 20th observation is 167 and the 21st is 168, so the median height = (167 + 168)/2 = 167.5.

For method (c), draw a horizontal line on Fig. 3.5 at a frequency of 20.5. Where this line meets the polygon, the corresponding height is the median height (see Fig. 4.1). This method which gives a median of about 167 cm. Minitab's median for our data is 167.5 (see Table 4.1).

The four estimates of the median height are approximately equal (as they should be!).

4.5   Sample Mode

The **sample mode** of a variable x is defined as follows:

$$\text{Sample mode is the value with the highest frequency} \qquad (4.2)$$

Example

The heights of five students are 183, 163, 152, 157, and 157 cm. The mode is 157 cm because it occurs twice, while the others occur only once.

Example

The heights of four students are 165, 173, 180, and 164 cm. Since each value occurs the same number of times, we can conclude that either there is no mode, or there are four modes. The fact that the mode may not be unique is one of its disadvantages.

Example

Given the heights of 40 students in Table 1.1, we can use either a dotplot (see Fig. 3.1) or the stem and leaf display (Fig. 3.2) to obtain a mode of 157 cm, which occurs five times. However, this is hardly a 'middle value.' The modal *group*, as opposed to the modal *value*, for the variable height is 164.5 to 169.5 (see Table 3.2 and/or Fig. 3.4). This is perhaps a more useful idea than simply quoting the mode. Note that Minitab does not give the mode.

Example

For categorical data, we cannot calculate either the mean or the median. The mode, on the other hand, may have some limited use. For example, in our sample of 40 students, if 13 are male and 27 are female, then the modal sex is female.

4.6   When to Use the Mean, Median, and Mode

In order to decide which of the three 'averages' to use in a particular case we need to consider the shape of the distribution as indicated by a graph such as a dotplot (see Fig. 3.1, for example), the histogram (see Fig. 3.4 for a continuous variable example), or the line chart (see Fig. 3.7 for a discrete variable example). For categorical data, the mode is the only one of the three averages which is defined.

If the shape of the distribution is roughly **symmetrical** about a vertical centre line, then the sample mean is the preferred average. Such is the case in Figs. 3.1 and 3.4, which are graphical plots for the heights of 40 students. You may have noticed that the mean and median heights for these data were almost identical, while the mode was not at all representative of the data:

sample mean   = 168.2 cm
sample median = 167.5 cm
sample mode   = 157 cm.

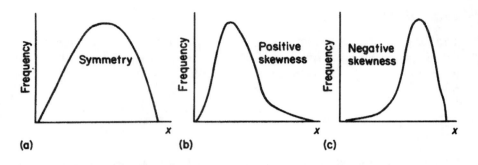

Figure 4.2 Symmetry and Skewness: (a) Mean = Median = Mode; (b) Mean > Median > Mode; and (c) Mean < Median < Mode

So why should the mean be preferred to the median in this case? The answer is a theoretical one, which you are asked to take on trust, namely, the sample mean is a more precise measurement for such distributions.

If the shape of the distribution is not symmetrical, it is described as **skew**. Fig. 4.2 shows three sketches of the 'shape' of three distributions exhibiting symmetry, positive skewness, or negative skewness, respectively. It also indicates the rankings of the mean, median, and mode in each of the three cases. For markedly skew data, there will be a small number of extremely high values (Fig. 4.2[b]) or low values (Fig. 4.2[c]), which are not balanced by values on the other side of the distribution. The sample mean is more influenced than the median by these extreme values. So the sample median is preferred for data showing marked skewness.

By 'marked skewness' we mean that the measure of skewness (see Section 4.13) is greater than 1 or less than −1, as a rough guide. If in doubt, both the sample mean and the sample median should be quoted.

The mode is not much use for either continuous or discrete data, since it may not be unique (as we saw in Section 4.5) or it may not exist at all, and for other theoretical reasons. The mode is useful only for categorical data.

Occasionally, distributions arise for which none of the three 'averages' is particularly informative.

Example

Table 4.2 shows the number of cigarettes smoked by 50 subjects. Drawing a dotplot (see Fig. 4.3) shows a positively skew distribution. The mean number of cigarettes smoked per day is equal to:

$$(0 \times 30 + 10 \times 10 + 20 \times 5 + 30 \times 3 + 4 \times 2)/50 = 7.4$$

**Table 4.2   The Number of Cigarettes
Smoked per day by 50 Subjects**

Number of Cigarettes	Number of Subjects
0	30
10	10
20	5
30	3
40	2

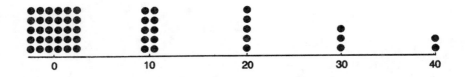

Figure 4.3   Dotplot of the Number of Cigarettes Smoked Per Day by 50 Subjects

However, this number does not seem to 'represent' the data very well. Neither does the median, which is zero (the mean of the 25th and the 26th value), nor the mode, which is also zero.

Table 4.2 and Fig. 4.3 are both very informative, but if we *must* summarise these data numerically, we could state that 60% of subjects are nonsmokers, while smokers smoke a mean of 18.5 cigarettes per day.

4.7   Measures of Variation

Averages are not the whole story. They do not give a complete description of a set of data and can, on their own, be misleading. The definition of a statistician as one who, on plunging one foot into a bucket of boiling water and the other in a bucket of melting ice, declares, "On average I feel just right!" completely misses the purpose of statistics, which is to collect and analyse data which *vary*.

However, it is not the aim of this book to lament the misconceptions some people have about statistics, but hopefully to educate and inform. So it would be more reasonable for the caricatured statistician to feel unhappy because the temperature of his feet varies so greatly about a comfortable average.

Similarly, an employee feels unhappy when told 'wages have risen by 10% in the past year' if his own wage has risen by only 3%, while the cost of living has risen by 8% (both the 8% and the 10% are averages, by the way).

Two measures of variation will be discussed in Sections 4.8, 4.9, and 4.10 in some detail, and three other measures of variation will be mentioned briefly in Section 4.12.

4.8 Sample Standard Deviation (s)

One way of measuring variation in sample data is to sum the differences between each observed value and the sample mean, $\bar{x}$, to give:

$$\Sigma(x - \bar{x})$$

However, this always gives the answer zero, as we saw in Section 2.2 and three times in Worksheet 2.

A more useful measure of variation, called the **sample standard deviation**, s, is obtained by summing the squares of the differences $(x - \bar{x})$, dividing by $n - 1$ (where n is the number of observations in the sample, more commonly known as the 'sample size'), and taking the square root. This gives a kind of 'root mean square deviation' (see the formula for s below).

The reason for squaring the differences is that this makes them positive or zero. The reason for dividing by $n - 1$ rather than n is discussed later in this section. The reason for taking the square root is to make the measure have the same units as the variable x. There are more theoretical reasons than these for using standard deviation as a measure of variation, but I hope the above will give you an intuitive feel for the formulae which are now introduced.

Sample standard deviation, s, may be defined by the formula:

$$s = \sqrt{\frac{\Sigma(x - \bar{x})^2}{n - 1}} \qquad (4.3)$$

An alternative form of this formula which is easier to use for calculation purposes is

$$s = \sqrt{\frac{\Sigma x^2 - \frac{(\Sigma x)^2}{n}}{n - 1}} \qquad (4.4)$$

Example

The heights of a sample of five people are 183, 163, 152, 157, and 157 cm. Therefore:

$$\Sigma x = 183 + 163 + 152 + 157 + 157 = 812$$
$$\Sigma x^2 = 183^2 + 163^2 + 152^2 + 157^2 + 157^2 = 132,460$$
$$n = 5$$

The sample standard deviation is

$$s = \sqrt{\frac{132,460 - \frac{812^2}{5}}{5 - 1}}$$

$$s = 12.2 \text{ cm}$$

Example

For the heights of a sample of 40 students given in Table 1.1, we can calculate that $\Sigma x = 6730$, $\Sigma x^2 = 1,135,558$, $n = 40$.

So the sample standard deviation is

$$s = \sqrt{\frac{1,135,558 - \frac{6730^2}{40}}{39}}$$

$$= 9.1 \text{ cm}$$

Notes

(a) The units of standard deviation are the same as the units of the variable height, i.e., centimetres, in both examples above.

(b) The answer should be given to one more significant figure than the raw data, i.e., to one decimal place in both the examples above.

A question which is often asked is, 'Why use $n - 1$ in the formulae for s?' The answer is that the values we obtain give better estimates of the standard deviation of the population than would be obtained if we had used n instead. In what is called 'Statistical Inference,' a major topic from Chapter 8 to the end of this book, we are not so much interested in sample data as we are in what conclusions, based on sample data, can be drawn about the population from which the sample was taken.

Another natural question at this stage is 'Now we have calculated the sample standard deviation, what does it tell us?'. The answer is 'Be patient!' When we have discussed the 'normal distribution' in Chapter 7, standard deviation will become more meaningful. For the moment please accept the basic idea that standard deviation is a measure of variation about the mean. The more variation there is in the data, the higher will be the standard deviation. If there is no variation at all, the standard deviation will be zero. It can never be negative.

For the height data in Table 1.1, we obtained a number of statistics using Minitab (see Table 4.1 in Section 4.3). The package gives a standard deviation (Stdev) of 9.11 cm, which agrees with the value found earlier using the formula.

4.9 Sample Inter-Quartile Range

Just as the sample median is such that half the observed values are less than it, and it is the $(n + 1)/2^{th}$ value, we define the lower and upper quartiles in a similar way.

The lower quartile, Q1, is such that one-quarter of the observed values are less than it, or formally:

Sample **lower quartile**, Q1, is the $(n + 1)/4^{th}$ value $\qquad$ (4.5)

Similarly, the upper quartile, Q3, is such that three-quarters of the observed values are less than it, or:

Sample **upper quartile**, Q3, is the $3(n + 1)/4^{th}$ value $\qquad$ (4.6)

The **sample inter-quartile range** is defined as the difference between the upper and lower quartiles, that is:

sample **inter-quartile range** = upper quartile − lower quartile
$$= Q3 - Q1 \qquad (4.7)$$

Example

The heights of a sample of five people are 183, 163, 152, 157, and 157 cm.
In rank order these are 152, 157, 157, 163, 183. Since $n = 5$,

$(n + 1)/4 = 1.5$, lower quartile $= 152 + 0.5 (157 - 152) = 154.5$.

$3(n + 1)/4 = 4.5$, upper quartile $= 163 + 0.5 (183 - 163) = 173$.

So, inter-quartile range $= 173 - 154.5 = 18.5$.

Example

To find the quartiles for the heights of a sample of 40 students given in Table 1.1, instead of ranking the 40 values, we could use one or more of the four methods we used to find the median (see Section 4.4). We refer to only two of these methods here:

(a) Dotplot (see Fig. 3.1). Since $n = 40$, $(n + 1)/4 = 10.25$; the 10th value is 161, the 11th is 163, so:
   lower quartile $= 161 + 0.25 (163 - 161) = 161.5$
   Similarly, $3(n + 1)/4 = 30.75$;
   the 30th value is 175, the 31st is 177, so:
   upper quartile $= 175 + 0.75 (177 - 175) = 176.5$.
   So, inter-quartile range $= 176.5 - 161.5 = 15$ cm.

(b) For the same data, Minitab gives exactly the same values for the lower and upper quartiles, which it refers to as Q1 and Q3, respectively. Q2 is, of course, the median. Minitab does not calculate the inter-quartile range. Notice that the middle half of the observed values lie between Q1 and Q3.

4.10 When to Use Standard Deviation and Inter-Quartile Range

In order to decide which of these two measures of variation to use in a particular case, the same considerations apply as for averages (refer to Section 4.6 if necessary). So, for roughly symmetrical data, use standard deviation. For markedly skew data, use inter-quartile range.

4.11 Box and Whisker Plots

These plots were mentioned in Section 3.2 (see Fig. 3.3) before the terms median, lower quartile, and upper quartile had been introduced. Fig. 3.3 is reproduced here as Fig. 4.4 for convenience. From left to right, the five values 152, 161.5, 167.5, 176.5, and 184, respectively, are the minimum (smallest), lower quartile (Q1), median, upper quartile (Q3), and maximum (largest) values in the sample.

For the height data, the five values of interest may be obtained without a box and whisker plot (see Table 4.1 in Section 4.3).

Also, we could compare the heights of, for example, male and female students by means of two box and whisker plots, using a similar method

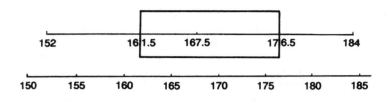

Figure 4.4   Box and Whisker Plot for the Data in Table 3.1

to that used in Section 3.6 in which two dotplots were drawn on the same scale. This is left as an exercise for the reader.

4.12   Other Measures of Variation

We will consider three other measures of variation briefly.

Variance is simply the square of the standard deviation, so we can use the symbol s^2. Variance is a common term in many statistical methods which involve what is called the 'analysis of variance' (ANOVA), most of which are beyond the scope of this book. However, Chapter 12 provides an introduction to ANOVA, while in Chapter 15 we see a particularly useful application to a topic called Regression.

Coefficient of variation is defined as $100s/\bar{x}$, and is expressed as a percentage. This is used to compare the variabilities of two sets of data when there is an obvious difference in magnitude in both the means and standard deviations. For example, to compare the variation in the heights of boys aged 5 and 15 years, suppose $\bar{x}_5 = 100$, $s_5 = 6$, $\bar{x}_{15} = 150$, $s_{15} = 9$, then both sets have a coefficient of variation of 6%.

Range is defined as the difference between the largest observed value and the smallest observed value, when we are discussing sample data. It is commonly used because it is easy to calculate, but it is unreliable except in special circumstances because only two of the sample observations are used to calculate it. Also the more sample observations we take, the larger the range is likely to be.

4.13   A Measure of Skewness

We saw in Section 4.6 that if the distribution of a set of data is perfectly symmetrical, then the mean and median are equal. If there is positive skewness, then the mean exceeds the median, while the mean is less than the median for negatively skew data. The following dimensionless measure of skewness is therefore zero, positive, or negative, depending on the type of skewness:

$$\text{Measure of skewness} = \frac{3 \; (\text{sample mean} - \text{sample median})}{\text{sample standard deviation}} \qquad (4.8)$$

As a rough guide, if this measure is greater than 1 we can say that the distribution is 'markedly positively skew'. If it is less than −1 we can conclude that the distribution is 'markedly negatively skew'. If the measure of skewness lies between −1 and +1, we can say that the distribution is roughly symmetrical.

Example

For the distribution of the heights of 40 students from Table 1.1, sample mean = 168.2, sample median = 167.5, sample standard deviation = 9.1, so:

$$\text{measure of skewness} = \frac{3(168.2 - 167.5)}{9.1} = 0.23$$

The distribution of heights shows slight positive skewness.

4.14 Summary

When a variable is measured for a number of individuals, the resulting data may be summarised by calculating averages and measures of variation. In addition, a measure of skewness is sometimes useful. The particular type of average and measure of variation required depends on the type of variable and the shape of the distribution. Some examples are given in Table 4.3. Three other measures of variation are the variance, coefficient of variation, and range.

Table 4.3 Examples of Averages and Measures of Variation

Type of Variable	Shape of Distibution	Average	Measure of Variation
Continuous or discrere	Roughly symmetrical, unimodal	Sample mean $(\bar{x})$	Sample standard deviation (s)
Continuous or discrete	Markedly skew, unimodal	Sample median	Sample inter−quartile range
Categorical		Sample mode	

Worksheet 4: Summarizing Data by Numerical Measures

Questions 1 to 7 are multiple choice. Choose one of the three options in each case.

1. The lower quartile of a distribution is such that:
   (a) 1/4 of the values are greater than it,
   (b) 1/4 of the values are less than it,
   (c) 3/4 of the values are less than it.
2. The standard deviation of the numbers 6, 7, and 8 is 1. If 1 is added to each number the standard deviation becomes: (a) 1, (b) 2, (c) $\sqrt{2}$.
3. The average which represents the value of a total when shared out equally is the: (a) mean, (b) median, (c) mode.
4. The mean of the numbers 6, 7, and 8 is 7. If each number is squared the mean becomes (a) 49, (b) greater than 49, (c) less than 49.
5. For a symmetrical distribution:
   (a) mode = median = mean,
   (b) mode > median > mean,
   (c) mode < median < mean.
6. A symmetrical distribution always has:
   (a) A bell shape,
   (b) A mean and a median of the same value,
   (c) No extremely high or low values.
7. Which summary statistics are preferred when the distribution is roughly symmetrical?
   (a) Median and inter-quartile range,
   (b) Mode and range,
   (c) Mean and standard deviation.
8. (a) Why do we need averages?
   (b) Which average can have more than one value?
   (c) Which average has the same number of observations above it as below it?
   (d) When is the sample median preferred to the sample mean?
   (e) When is the sample mode preferred to the sample mean?
   (f) When is the sample mean preferred to both the sample median and the sample mode?
9. (a) Why do we need measures of variation?
   (b) What measure of variation is most useful in the case of (i) a symmetrical distribution, (ii) a skew distribution?
   (c) Think of an example of sample data where the range would be a misleading measure of variation.

(d) Name the measure of variation associated with the (i) sample mean, (ii) sample median, (iii) sample mode.

(e) Name the average associated with the (i) sample standard deviation, (ii) sample inter-quartile range, (iii) range.

10. The weekly incomes (£) of a random sample of part-time window cleaners are 75, 67, 60, 62, 65, 67, 62, 68, 82, 67, 62, and 200.

(a) Find the sample mean, sample median, and sample mode of weekly income. Why are your three answers different?

(b) Find the sample standard deviation and the sample inter-quartile range. Why are your answers different?

(c) Which of the measures you have obtained are the most useful in summarizing the data?

Try this question by hand calculation, and check your answers using Minitab.

11. Eleven cartons of sugar, each nominally containing 1 kg, yielded the following weights of sugar 1.02, 1.05, 1.08, 1.03, 1.00, 1.06, 1.08, 1.01, 1.04, 1.07, and 1.00. Calculate the sample mean and sample standard deviation of the weight of sugar. Try this question by calculator and by Minitab.

12. Using the data in Question 2 of Worksheet 3, find:

(a) The sample mean and standard deviation,

(b) The sample median and inter-quartile range.

Decide which is the preferred (i) average and (ii) measure of variation.

13. For the distance data in Table 1.1, find:

(a) The sample mean and standard deviation,

(b) The sample median and inter-quartile range.

Decide which is the preferred (i) average and (ii) measure of variation.

14. For the distance data in Table 1.1, compare the distances of male and female students graphically using appropriate numerical measures.

15. Consider again Question 5 of Worksheet 3. Having read Chapter 4, discuss the relative merits of using the mean, median, and mode to obtain the 'average' number of goals per team. Combine the data from all four divisions to answer this question.

Chapter 5

Probability

Dr. Price estimates the chance in favour of the wife being the survivor in marriage as 3 to 2.

5.1   Introduction

The preceding chapters of this book have been concerned with statistical data and methods of summarising such data. We can think of such sample data as having been drawn from a larger 'parent' population. Conclusions from sample data about populations (which is a branch of statistics called 'statistical inference', see Chapter 8 onwards) must necessarily be subject to some uncertainty since the sample cannot contain all the information in the population. This is one of the main reasons why **probability**, which **is a measure of uncertainty**, is now discussed.

Probability is a topic which may worry you, either because you have never studied it before, or you have studied it before but you did not fully get to grips with it. It is true that the study of probability requires a clear head, a logical approach, and the ability to list all the outcomes of simple experiments, often with the aid of diagrams. After some experience and some (possibly painful) mistakes, which are all part of the learning process, the penny usually begins to drop.

Think about the following question which will give you some feedback on your present attitude towards probability (do **not** read the discussion until you have thought of an answer).

Probability Example 5.1

A person tosses a coin five times. Each time it comes down heads. What is the probability that it will come heads on the sixth toss?

Discussion

If your answer is '1/2' (or 'a half' or '1 in 2' or '50%'), you are assuming that the coin is 'fair', meaning that it is equally likely to come down heads or tails. You have ignored the 'statistical data' that all five tosses resulted in heads.

If your answer is 'less than 1/2' you may be quoting 'the law of averages' which presumably implies that, in the long run, half the tosses will result in heads and half in tails. This again assumes that the coin is fair. Also, do six tosses constitute a long run of tosses, and does the 'law of averages' apply to each individual toss?

If your answer is 'greater than 1/2', perhaps you suspect that the coin has two heads, in which case the probability of heads would be 1, or that the coin has a bias in favour of heads. Think about this teasing question again when you have read this chapter.

5.2 Basic Ideas of Probability

One dictionary definition of probability is 'the extent to which an event is likely to occur, measured by the ratio of the favourable cases to the whole number of cases possible'. Consider the following example.

Probability Example 5.2

A ball is selected at random from a bag containing three red balls and seven white balls. The probability that a red ball will be drawn is 3/10. Note the following points:

(a) **'At random'** means that each of the 10 balls has the same chance (probability?) of being selected, implying that we mix up the balls and the person selecting the ball should look away or close his/her eyes. We say that the 10 outcomes are 'equally likely' in this case.

(b) **Probability is a measure of uncertainty** which, as we shall see later, can take any value between 0 and 1.

(c) The probability that a white ball will be drawn is 7/10. Note that the total of the two probabilities is 3/10 + 7/10 = 1, and that no other outcome is possible.

Does Fig. 5.1 help you to understand Example 5.2? It helps *me* to visualise a probability problem either in my head or on paper, and the

**Figure 5.1   A Bag Containing Three
Red and Seven White Balls**

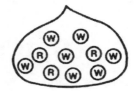

more complex the problem the more useful a visual aid is likely to be, as we shall see later in this chapter.

Recalling the dictionary definition of probability at the beginning of this section, the number of favourable cases is 3 for the red ball event, out of a total of 10 possible cases, and the required probability is again 3/10.

In order to gain an understanding of probability, it is helpful to define three terms which have a special meaning when we discuss probability. The terms are

Trial, Experiment, and Event. The definitions are

A **Trial** is an action which results in one of several possible outcomes.
An **Experiment** is a series of trials (or possibly just one).
An **Event** is a set of outcomes with something in common.

In Example 5.2 above,

The trial is 'drawing a ball from a bag'.
The experiment is also 'drawing a ball from a bag', since only one ball is selected.
The event is 'red ball', corresponding to 3 of the 10 possible outcomes.

5.3   The *a priori* Definition of Probability for Equally Likely Outcomes

This section is a more formal look at a definition of probability for experiments whose outcomes are equally likely, as in Example 5.2. Suppose each trial in an experiment can result in one of n 'equally likely' outcomes, r of which correspond to an event, E. Then the probability of event E is r/n, which we write:

$$P(E) = \frac{r}{n} \qquad (5.1)$$

This *a priori* definition has been used for Example 5.2; event E is 'red ball', $n = 10$ since it is assumed that each of the 10 balls is equally likely to be drawn from the bag, and $r = 3$ since 3 of the 10 balls are red and therefore correspond to the event E. So we write

$$P(\text{red ball}) = \frac{3}{10}$$

Note the following points:

(a) We only have to think about the possible outcomes, we do not actually have to carry out an experiment of removing balls from a bag. The Latin phrase *a priori* means 'without investigation or sensory experience'.

(b) It is necessary to know that the possible outcomes are equally likely to occur. This is why this definition is called a 'circular' definition, since equally likely and equally probable have the same meaning. More importantly, we should not use the *a priori* definition if we do not know that the possible outcomes are equally likely. (Example: 'Either I will become the manager of the England soccer team or I will not, so the probability that I will is $1/2$, and the same probability applies to everybody'. This is clearly an absurd argument.)

The *a priori* definition is most useful in games of chance.

5.4 The Relative Frequency Definition of Probability, Based on Experimental Data

Probability Example 5.3

If an ordinary drawing pin is tossed in the air, it can fall in one of two ways: with the point upwards, which we shall call event U, or point downwards, which we will call event U'. (We will assume that no other event, such as the drawing pin balancing on its point, is possible.) We cannot obtain an *a priori* estimate of the probability of event, U, i.e., P(U), but we can estimate this probability by carrying out an experiment as follows.

Toss the drawing pin 50 times and record the result of each of the 50 trials as U or U'. Suppose that 28 U occurred, then our estimate of P(U) is simply $28/50 = 0.56$.

Formally, the relative frequency definition of probability is as follows. If, in a large number of independent trials, n, r of these trials result in event E, the probability of event E is $\frac{r}{n}$. So we write:

$$P(E) = \frac{r}{n} \tag{5.2}$$

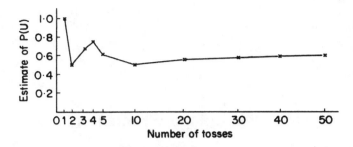

Figure 5.2   Estimating the Probability That a Drawing Pin Will Fall Point Upwards

Notes

(a) The number of trials, n, must be large. The larger the value of n, the better is the estimate of the probability. (How large is large? The only satisfactory answer at this stage is 'as large as   practicable'. We will have a 'better' answer in Chapter 9.)

(b) The word 'independent' here means that the outcome of any of the 50 tosses does not depend on the results of previous tosses, i.e., no cheating! (I suggest holding the drawing pin in a cupped upturned hand, and throw it   3 feet, say, up in the air and allow it to fall on the floor or carpet. Immediately note the result. Repeat a total of 50 times.)

(c) One theoretical problem with the relative frequency definition of probability is that there is no guarantee that the value of r/n will settle down to a constant value as the number of trials gets larger and larger. However, if you estimate P(U) after 1, 2, 3, 4, 5, 10, 20, 30, 40, and 50 tosses, the graph of P(U) against the number of trials gives the impression that it is settling down (see Fig. 5.2) to about 0.58.

5.5   The Range of Possible Values for a Probability Value

Using either of the two definitions of probability, we can show that probabilities can only take values between 0 and 1. The value of r must take one of the integer values between 0 and n, *so* r/n can take values between $0/n$ and n/n, that is 0 and 1.

If $r = 0$, we are thinking of an event which cannot occur (*a priori* definition) or an event which has not occurred in a large number of trials (relative frequency definition). For example, the probability that I will throw a 7 with one ordinary die is 0.

If $r = n$, we are thinking of an event which must always occur (*a priori* definition) or an event which has occurred in each of a large number of trials (relative frequency definition). For example, the probability that the sun will rise tomorrow can be assumed to be 1, unless you are a pessimist (see Section 5.7).

5.6 Probability, Percentage, Proportion, and Odds

We can convert a probability to a percentage by multiplying it by 100. So a probability of 3/4 implies a percentage of 75%.

We can also think of probability as meaning the same thing as proportion. So a probability of 3/4 implies that the proportion of times an event will occur is also 3/4.

A probability of 3/4 is equivalent to odds of 3/4 to 1/4, which is usually expressed as 3 to 1.

5.7 Subjective Probability

There are other definitions of probability apart from the two discussed earlier in this chapter. We all use **'subjective probability'** in forecasting future events, for example, when we try to decide whether it will rain tomorrow, and when we try to assess the reactions of others to our opinions and actions. We may not be quite so calculating as to estimate a probability value, but we may regard future events as being probable, rather than just possible. In subjective assessments of probability we may take into account experimental data from past events, but we are likely to add a dose of subjectivity depending on our personality, our mood, and other factors.

5.8 Probabilities Involving More Than One Event

Suppose that we are interested in the probabilities of two possible events, E_1 and E_2. For example, we may wish to know the probability that both events will occur, or perhaps the probability that either or both events will occur. We will refer to these as, respectively,

$$P(E_1 \text{ and } E_2) \quad \text{and} \quad P(E_1 \text{ or } E_2 \text{ or both}).$$

In set theory notation these compound events are called the **intersection** and **union** of events E_1 and E_2, and their probabilities are written:

$$P(E_1 \cap E_2) \quad \text{and} \quad P(E_1 \cup E_2)$$

There are two probability laws which can be used to estimate such probabilities, and these are discussed in Sections 5.9 and 5.10.

5.9 Multiplication Law (The 'and' Law)

The general case of the **multiplication** law is

$$P(E_1 \text{ and } E_2) = P(E_1)P(E_2|E_1) \tag{5.3}$$

where $P(E_2|E_1)$ means the probability that event E_2 will occur, given that event E_1 has already occurred. The vertical line between E_1 and E_2 should be read as 'given that' or 'on the condition that'.

$P(E_2|E_1)$ is an example of what is called a **conditional** probability.

Probability Example 5.4

If two cards are selected at random, one at a time **without replacement** from a pack of 52 playing cards, what is the probability that both cards will be aces?

P(two aces) = P(first card is ace and second card is ace), which is logical.

= P(first card is ace) × P(second card is ace | first card is ace),

using the multiplication law, where E_1 = first card is ace, E_2 = second card is ace = $\frac{4}{52} \times \frac{3}{51}$ (see Fig. 5.3) = 0.0045. (Four decimal places are usually more than sufficient for a probability value.)

In many practical examples the probability of event E_2 does not depend on whether E_1 has occurred. In this case we say that events E_1 and E_2 are **statistically independent**, which is often shortened to **independent**, giving rise to the special case of the multiplication law:

$$P(E_1 \text{ and } E_2) = P(E_1)P(E_2) \tag{5.4}$$

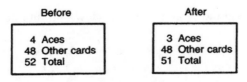

Figure 5.3   Before and After the First Card is Drawn, Without Replacement

Figire 5.4   Before and After the First Card is Drawn, With Replacement

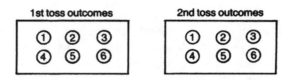

Figure 5.5   Tossing a Die Twice

Probability Example 5.5

Consider the previous Example (5.4), and change the phrase 'without replacement', to **'with replacement'**.

P(two aces)   = P(first card is ace and second card is ace), as in Example 5.4,

   = P(first card is ace) × P(second card is ace),

using the multiplication law for independent events, $= \frac{4}{52} \times \frac{4}{52} = 0.0059$.

Because the first card is replaced, the probability that the second card is an ace will be 4/52 whatever the first card is.

Notes

(a) Comparing the two cases of the multiplication law we can state that, if two events E_1 and E_2 are statistically independent, then

$$P(E_2 | E_1) = P(E_2)$$

(b) Clearly we could write the general law of multiplication alternatively as:

$$P(E_1 \text{ and } E_2) = P(E_2)P(E_1 | E_2)$$

by swapping E_1 and E_2.

5.10 Addition Law (The 'or' Law)

The general case of the **addition law** is

$$P(E_1 \text{ or } E_2 \text{ or both}) = P(E_1) + P(E_2) - P(E_1 \text{ and } E_2) \qquad (5.5)$$

Probability Example 5.6

If a die is tossed twice, what is the probability of getting at least one 5? Here, defining E_1 as the event '5 on first toss', and E_2 as the event '5 on second toss',

P(at least one 5) = P(5 on first toss or 5 on second toss or 5 on both tosses),

= P(5 on first toss) + P(5 on second toss) −P(5 on both tosses)

$= \frac{1}{6} + \frac{1}{6} - \frac{1}{6} \times \frac{1}{6}$

$= 0.3056$

Here we have used the addition law, followed by the multiplication law for independent events.

In many practical cases the events E_1 and E_2 are such that they cannot both occur. In this case we say that the two events are **mutually exclusive**, giving rise to the special case of the addition law:

$$P(E_1 \text{ or } E_2) = P(E_1) + P(E_2) \qquad (5.6)$$

for mutually exclusive events E_1 and E_2.

Note: comparing the versions of the addition law, we see that if E_1 and E_2 are mutually exclusive, then $P(E_1 \text{ and } E_2) = 0$.

Probability Example 5.7

If a die is tossed twice, what is the probability that the total score is 11? There are only two ways of getting a total of 11, either by getting 6 on the first toss and 5 on the second toss, or by getting 5 on the first toss and 6 on the second toss. These two ways, which we will call E_1 and E_2, are mutually exclusive.

P(total *of* 11) = $P(E_1) + P(E_2)$

$= \frac{1}{6} \times \frac{1}{6} + \frac{1}{6} \times \frac{1}{6}$

$= 0.0556$

We could have obtained the answer to Example 5.7 by noting, from Fig. 5.6, that 2 of the 36 equally likely cases (each has a probability of

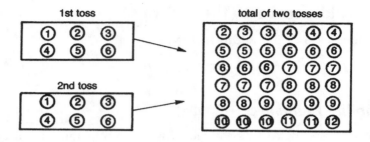

Figure 5.6   The Total Score When a Die is Tossed Twice

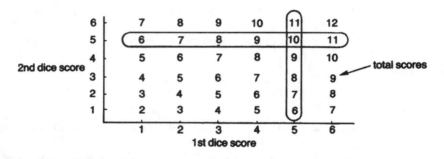

Figure 5.7   The Total Score When Two Dice Are Thrown

1/36) results in a total of 11. Hence, $P(11) = \frac{2}{36} = 0.0556$, using the *a priori* definition of probability.

Similarly the answer to Probability Example 5.6 could be obtained by using the display shown in Fig. 5.7.

P(5 on first toss or 5 on second toss or both) = 11/36 = 0.3056, as before, since there are 11 favourable cases out of 36, and all 36 are equally likely.

5.11   Mutually Exclusive and Exhaustive Events

If all possible outcomes of an experiment are formed into a set of mutually exclusive events, we say they form a **mutually exclusive and exhaustive** set of events which we will call $E_1, E_2, ..., E_n$, if there are n events. Applying the special case of the law of addition,

$$P(E_1 \text{ or } E_2 \text{ or} ... \text{or } E_n) = P(E_1) + P(E_2) + ... + P(E_n)$$

But since the events are exhaustive, one of them must occur, and so the left-hand side of the equation is 1. In other words, then,

The sum of the probabilities of a set of mutually exclusive and exhaustive events is 1.

This result is useful in checking whether we have correctly calculated the separate probabilities of the various mutually exclusive events of an experiment.

Probability Example 5.8

For families with two children, what are the probabilities of the various possibilities, assuming that boys and girls are equally likely at each birth? Four mutually exclusive and exhaustive events are BB, BG, GB, and GG, where BG means a boy followed by a girl. Therefore,

$$P(BG) = \frac{1}{2} \times \frac{1}{2} = \frac{1}{4}$$

using the special law of multiplication. Similarly, $P(BB) = P(GB) = P(GG) = \frac{1}{4}$, must occur, and the total probability is 1.

5.12 Complementary Events and the Calculation of P (at Least 1 …)

For any event E, there is a **complementary** event E' which we call *not* E. Since either E or E' must occur, and they cannot both occur,

$$P(E) + P(E') = 1,$$

which is a special case of the result of the previous section. It follows that:

$$P(E) = 1 - P(E').$$

This result is useful in some cases, where it is easier to calculate the probability of the **complement** of some event and subtract the answer from 1 than it is to calculate the probability of the event directly. This is especially true when we wish to calculate the probability that 'at least 1 of something will occur in a number of trials', since:

$$P(\text{at least } 1\ldots) = 1 - P(\text{none}\ldots) \tag{5.7}$$

Probability Example 5.9

For families with four children, what is the probability that there will be at least one boy, assuming boys and girls are equally likely? Instead of listing the 16 outcomes BBBB, BBBG, etc., we simply say:

$$P(\text{at least 1 boy}) = 1 - P(\text{no boys})$$
$$= 1 - P(GGGG)$$
$$= 1 - \frac{1}{2} \times \frac{1}{2} \times \frac{1}{2} \times \frac{1}{2}$$
$$= \frac{15}{16}$$

5.13   Probability Trees

A **probability tree** (sometimes called a tree diagram) can be used instead of the laws of probability when we are considering the outcomes of an experiment consisting of a sequence of a few trials.

Probability Example 5.10

If two cards are selected at random, one at a time without replacement, from a pack of 52 playing cards, investigate the probability of the events: *ace(A)* and *not ace (A')* (see Fig. 5.8).

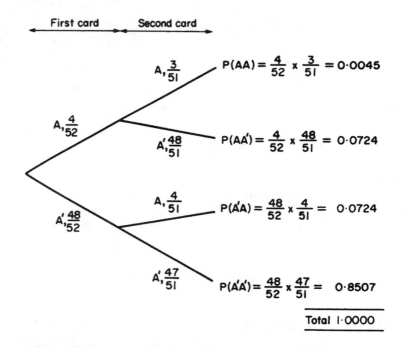

Figure 5.8   Probability Tree for Two Cards, Without Replacement

In general, in a probability tree, the events resulting from the first trial are represented by two or more branches from a starting point. More branches are added to represent the events resulting from the second and subsequent trials. The branches are labelled with the names of the events and their probabilities, taking into account previous branches (so we might be dealing with conditional probabilities as in Example 5.4). In order to calculate the probabilities of the set of mutually exclusive and exhaustive events, corresponding to all the ways of getting from the starting point to the end of each branch, probabilities are multiplied (so we are really using the law of multiplication). The total probabilty should, of course, equal 1.

For the above example:

P(two aces) $= P(AA') = 0.0045$
P(one ace) $= P(AA'$ or $A'A) = P(AA') + P(A'A) = 0.1448$
P(no aces) $= P(A'A') = 0.8507$

In general, the total number of branches needed will depend on both the number of possible events resulting from each trial and the number of trials. For example, if each of four trials can result in one of two possible outcomes there will be $2 + 4 + 8 = 16 = 30$ branches, or if each of three trials can result in one of three possible outcomes there will be $3 + 9 + 27 = 39$ branches.

Clearly, it becomes impractical to draw a probability tree if the number of branches is large. In the next chapter we will discuss more powerful ways of dealing with any number of independent trials, each with only two possible outcomes.

5.14 Venn Diagrams and Rees Diagrams

Venn diagrams can be used to help us to disentangle probability problems. Strictly speaking, a Venn diagram is used to describe 'sets', represented by areas within a larger area, and mutually exclusive 'elements' within sets, e.g., Fig. 5.9.

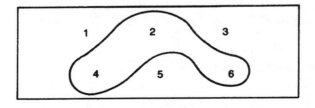

Figure 5.9 **Venn Diagram for the Six Elements When a Die is Thrown, and the Set (2, 4, 6)**

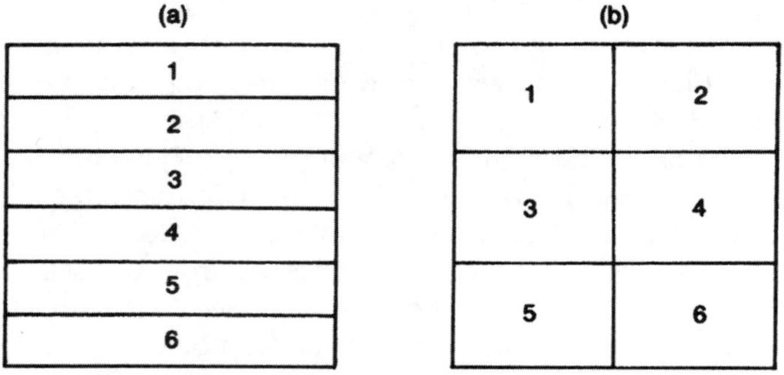

Figure 5.10   Rees Diagrams for the Outcomes of the Throw of a Die

If we take this a stage further and make the area corresponding to an element equal to the probability of that element, and make the total area equal to (the total probability of) 1, we have a modified diagram, which I will call a **Rees diagram** unless someone else has claimed it already! See Fig. 5.10 (a) and (b).

In both (a) and (b) of Fig. 5.10, the total area of the large rectangle equals 1 and this is subdivided into six equal areas of 1/6 — the probability of each outcome. Can this idea be used in more complicated examples ? Yes, see Fig. 5.11!

5.15   Summary

Probability as a measure of uncertainty may be defined using the *a priori* and relative frequency definitions. The first is useful in games of chance, the second when we have sufficient experimental data.

In calculating probabilities involving more than one event, two laws of probability are useful:

1. The multiplication law: $P(E_1 \text{ and } E_2) = P(E_1)P(E_2|E_1)$, which reduces to $P(E_1 \text{ and } E_2) = P(E_1)P(E_2)$, for statistically independent events.
2. The addition law: $P(E_1 \text{ or } E_2 \text{ or both}) = P(E_1) + P(E_2) - P(E_1 \text{ and } E_2)$, which reduces to $P(E_1 \text{ or } E_2) = P(E_1) + P(E_2)$, for mutually exclusive events.

Various graphical methods can be very helpful in representing the outcomes and their associated probabilities, for small experiments. Of these diagrams, the probability tree is one of the most helpful.

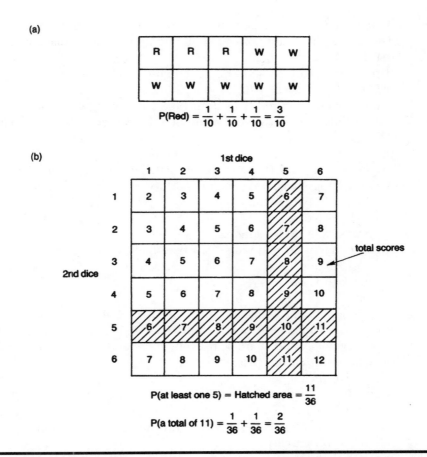

(a)

$$P(Red) = \frac{1}{10} + \frac{1}{10} + \frac{1}{10} = \frac{3}{10}$$

(b)

$$P(\text{at least one } 5) = \text{Hatched area} = \frac{11}{36}$$

$$P(\text{a total of } 11) = \frac{1}{36} + \frac{1}{36} = \frac{2}{36}$$

Figure 5.11   **Rees Diagrams Referring to Earlier Examples, So (a) Refers to Example 5.7 and (b) Refers to Example 5.8**

Worksheet 5:   Probability

Questions 1 to 10 are multiple choice. Choose one of the three options in each case.

1. Three cards are drawn without replacement from a well-shuffled pack. The probability that they are all diamonds is (a) 1/64, (b) 33/2704, (c) 11/850.
2. $P(B|A)$ means:
    (a) The probability of B divided by the probability of A,
    (b) The probability of B given that A has occurred,
    (c) The probability of A given that B has occurred.

3. A box contains 10 balls of which 5 are red and 5 are white. The probability that two white balls are drawn with replacement is (a) 1, (b) 0.25, (c) 0.20.

4. If three coins are tossed, the probability of two heads is (a) 3/8, (b) 2/3, (c) 1/8.

5. A bag contains six red balls, four blue balls, and two yellow balls. If two balls are drawn out without replacement, the probability that one ball will be red and the other will be blue is (a) 0.364, (b) 0.333, (c) 0.182.

6. Two events A and B are such that if B occurs, the probability of A is unchanged. The events are said to be
   (a) Mutually exclusive,
   (b) Exhaustive,
   (c) Statistically independent.

7. A bag contains six red balls, four blue balls, and two yellow balls. If two balls are drawn out with replacement, the probability that neither ball is red is (a) 0.227, (b) 0.250, (c) 0.750.

8. Two independent events A and B have probabilities $P(A) = 1/3$, $P(B) = 1/4$. Hence, P(A or B or both) is (a) 0.583, (b) 0.286, (c) 0.50.

9. If two events A and B are statistically independent, the occurrence of A implies that the probability of B occurring will be (a) 0, (b) unchanged, (c) 1.

10. A and B are mutually exclusive events, and $P(A) = 0.25$, $P(B) = 1/3$. P(A or B) is (a) 0.583, (b) 0.5, (c) 0.083.

11. Distinguish between the *a priori* and the relative frequency definitions of probability.

12. If the probability of a successful outcome of an experiment is 0.2, what is the probability of failure?

13. When two coins are tossed the result can be two heads, one head and one tail, or two tails, and hence each of these events has a probability of 1/3. What is wrong with this argument? What is the correct argument?

14. A coin is tossed five times. Each time it comes down heads. Hence the probability of heads is $5/5 = 1$. Discuss.

15. Three ordinary dice, one yellow, one blue, and one green, are placed in a bag. A trial involves selecting one die at random from the bag and rolling it, the colour and score being noted.
    (a) What does 'at random' mean here?
    (b) Write down the set of all possible outcomes.
    (c) Are the outcomes equally likely?
    (d) What is the probability of each outcome?
    (e) What are the probabilities of the following events:

      (i)   Yellow with any score?

      (ii) Yellow with an even score?

      (iii) Even score with any colour?

      (iv) Yellow 1 or blue 2 or green 3 ?

      (v) Neither even blue nor odd yellow ?

16. For the 27 female students whose heights are listed in Table 1.1, draw a histogram like Fig. 3.4. If one female student is selected at random, what is the probability that her height will be

    (a) Between 164.5 and 169.5 cm?

    (b) Between 149.5 and 179.5 cm?

    Express your answer to (a) as the ratio of two areas of your histogram.

17. The card game Patience (also known as Solitaire) may be defined as 'a game for one player in which cards taken from a well-shuffled pack have to be arranged in certain groups and sequences'. A player 'wins' a game if he/she finishes with four piles of cards, one for each suit, each in ascending order from ace, 2, 3, ...queen, king. But what is the probability of winning? If you have ever played Patience, you will know that you lose more often than you win, so the probability of winning is less than 0.5. The only way to get a better estimate of this probability is by playing the game a large number of times, and using the relative frequency definition of probability. Here are the results of 500 games, summarized in 10 blocks of 50 games (per block).

Game Numbers	Number of Wins	Cumulative Number of Wins
1 to 50	12	12
51 to 100	21	33
101 to 150	17	50
151 to 200	15	65
201 to 250	21	86
251 to 300	21	107
301 to 350	15	122
351 to 400	18	140
401 to 450	15	155
451 to 500	17	172

Estimate the probability of winning after each block of 50 games, and draw a graph of this probability vs. the number of games played (so far).

(a) Is 500 large enough, i.e., is the estimate of probability of winning settling down after 500 games?

(b) Are the games independent of each other?

18. Write down the following events in symbol form, where A and B are two events: (a) not A, (b) A given B, (c) B given A.

19. What is meant by: (a) P(A|B), (b) P(B|A), (c) P(A'), (d) A and B are statistically independent, (e) A and B are mutually exclusive? For (d) and (e), think of examples.

20. What is the 'and' law of probability, as applied to events A and B? What happens if A and B are statistically independent?

21. What is the 'or' law of probability, as applied to events A and B? What happens if A and B are mutually exhaustive?

22. What can be concluded if (a) P(A|B) = P(A), (b) P(A and B) = 0?

23. What is the probability of a 3 or a 6 with one throw of a die?

24. What is the probability of a red card, a picture card (ace, king, queen, or jack), or both, when a card is drawn from a pack at random?

25. A coin is tossed three times. Before each toss a subject guesses the result as heads or tails. If the subject always guesses tails, what is the probability that the subject will be correct: (a) three times, (b) twice, (c) once, (d) no times? Hint: draw a probability tree.

26. Three marksmen have probabilities 1/2, 1/3, and 1/4 of hitting a target with each shot. If all three marksmen fire simultaneously, calculate the probability that at least one will hit the target.

27. Of the sparking plugs manufactured by a firm, 3% are defective. In a random sample of four plugs, what is the probability that exactly one will be defective?

28. Suppose that, of a group of people, 30% own both a house and a car, 40% own a house, and 70% own a car. What proportion (a) own at least a house or a car, (b) of car owners are also householders?

29. Of 14 double-bedded rooms in a hotel, 9 have a bathroom. Of six single-bedded rooms, two have a bathroom.

(a) What is the probability that, if a room is randomly selected, it will have a bathroom?

(b) If a room is selected from those with a bathroom, what is the probability that it will be a single room?

30. A two-stage rocket is to be launched on a space mission. The probability that the lift-off will be a failure is 0.1. If the lift-off is successful the probability that the separation of the stages will be a failure is 0.05. If the separation is successful, the probability that the second stage will fail to complete the mission is 0.03.

What is the probability that the whole mission will: (a) be a success, (b) be a failure?

31. If one student is selected at random from the 40 listed in Table 1.1, what is the probability that this student is

(a) Male?

(b) Female?

(c) At least 165 cm in height?

(d) At least 165 cm in height, given that the student is
(i) male, (ii) female?

(e) Male, given that the student is (i) at least 165 cm in height, (ii) less than 165 cm in height? Do you think that sex is independent of height?

(f) Male and studying for a BSc?

(g) Male, or studying for a BSc, or both male and studying for a BSc?

32. This question is about "The Paradox of the Chevalier De Méré". He was a French nobleman in the 17th Century who was interested in the probabilities of two compound events.

The first was 'The probability of obtaining at least one 6 when a die is rolled 4 times'. The second was 'The probability of obtaining at least one double-6 when two dice are rolled 24 times'.

He thought that the two compound events had the same probability, namely, 2/3, presumably using 4/6 and 24/36, respectively. By calculating the correct probability values, show that he was wrong in both cases. (Use Section 5.12?)

33. This question is about the dice game called Craps (which was featured in the Hollywood musical "Guys and Dolls"):

(a) Write down the possible values of the total score, S, when a pair of fair die is thrown once, giving in each case the corresponding probabilities. What is the probability that, in a single throw of the two dice that one of the results $S = 7$ or $S = 11$ is obtained?

(b) Now suppose that, instead of being thrown only once, the pair of dice is thrown repeatedly until one of the results $S = 4$ or $S = 7$ is obtained (all other scores being disregarded). Show that the probability that $S = 4$ occurs before $S = 7$ occurs is equal to 1/3.

(c) Find the probabilities that, in repeated throwing (i) $S = 5$ occurs before $S = 7$, (ii) $S = 6$ occurs before $S = 7$.

(d) In the game of Craps, the person throwing the dice wins on the first throw if S is either 7 or 11, and loses if S is either 2, or 3, or 12. For any other value, k, of this first throw he must then throw both dice repeatedly and he wins provided $S = k$ occurs before $S = 7$. By drawing a suitable tree diagram, or otherwise, show that his total probability of winning is slightly less than 0.5. You may quote the result that, if x is a positive integer less than 1, $1 + x + x^2 + x^3 + \ldots = \frac{1}{1-x}$.

Chapter 6

Discrete Probability Distributions

6.1 Introduction

If a discrete variable can take values with associated probabilities, it is called a **discrete random variable** (r.v.). The values and the probabilities are said to form a **discrete probability distribution**. As a simple example, suppose we toss a fair coin once. The possible outcomes are heads and tails, so we will let the number of heads be our discrete random variable (we could equally well have chosen the number of tails). This variable can take the value 1, with probability 0.5, or 0, also with probability 0.5. In a table:

Table 6.1 Probability Distribution for the Number of Heads When a Coin is Tossed Once

Number of heads	0	1
Probability	0.5	0.5

Generalising, the number of possible values for any discrete r.v. must be an integer greater than or equal to 2. Also, the sum of all the corresponding probabilities must be equal to 1, but the probabilities need not be equal.

In this chapter we will study three standard discrete probability distributions, namely the Bernoulli, the binomial, and the Poisson. One other, the geometric distribution, will also be briefly introduced.

6.2 Bernoulli Distribution

A **Bernoulli trial** is defined as an action that results in one of two outcomes. Suppose that these outcomes, which are usually referred to as 'success' and 'failure' have probabilities p and (1 − p), respectively. Then the variable 'number of successes in a Bernoulli trial' is said to have: '**a Bernoulli distribution with parameter p**'.

This distribution can be set out in a table:

TABLE 6.2   Number of Successes in a Bernoulli Trial

Number of successes	0	1
Probabilities	(1 − p)	p

Comparing Tables 6.1 and 6.2, it is clear that the variable 'number of heads when a coin is tossed once, has a Bernoulli distribution with parameter 0.5.

Another way of expressing a Bernoulli distribution is in the form of a probability function, $P(x)$, as follows:

$$P(x) = p^x (1 - p)^{1-x} \quad \text{for} \quad x = 0, 1 \tag{6.1}$$

In Formula (6.1), $P(x)$ means the probability that there will be x successes in a Bernoulli trial. Clearly, x can be either 0 or 1. If we substitute, in turn, the values 0 and 1 in Formula (6.1), we obtain: $P(0) = 1 - p$ and $P(1) = p$, which agrees with the information in Table 6.2.

Finally, in this section we quote without proof that the mean and standard deviation of a variable which has a Bernoulli distribution are as follows:

$$\text{mean} = p \quad \text{standard deviation} = \sqrt{p(1 - p)} \tag{6.2}$$

6.3 Binomial Distribution

If we carry out a series of n independent Bernoulli trials, then the variable number of successes in the n trials is said to have: **a binomial distribution with parameters n and p**, or simply, a $B(n, p)$ distribution.

We will refer to such a series of trials as a **binomial experiment**. In order to decide whether a particular variable has a binomial distribution, we must check the following four conditions (which follow from the earlier part of this chapter):

1. There must be a fixed number of trials, n.
2. Each trial can result in one of only two outcomes, which we refer to as success and failure.
3. The probability of success in a single trial, p, is constant.
4. The trials are independent; in other words the probability of success in any one trial is unaffected by the result of any previous trial.

The probabilities of the various outcomes of a binomial experiment may be expressed in tabular form like Tables 6.1 or 6.2, but a more compact way is to quote the probability function $P(x)$ for any binomial distribution:

$$P(x) = \binom{n}{x} p^x (1 - p)^{n - x} \quad \text{for} \quad x = 0, 1, 2, \ldots n \qquad (6.3)$$

This formula is not difficult to use if each part is understood separately:
$P(x)$ means the probability of x successes in n trials.
$\binom{n}{x}$ is a shorthand for $\frac{n!}{x!(n - x)!}$ (refer to Section 2.3, if necessary).
$x = 0, 1, 2, \ldots, n$ means that we can use this formula for each of these values of x, which are the possible numbers of successes in n trials.

6.4 Calculating Binomial Probabilities: An Example

Suppose we toss a fair coin three times and we are interested in the number of heads we might obtain in three tosses, together with their corresponding probabilities. This example is similar to Question 25 of Worksheet 5, but we will answer it using a binomial formula, that is, if the four conditions are satisfied. Well are they?

1. Using the idea that a trial is a toss of the coin in this example, it follows that n is fixed at the value 3, i.e., $n = 3$.
2. There are only two possible outcomes when a coin is tossed, namely heads and tails. Since we are interested in the number of heads (as stated above), we will call heads a success, and then tails is a failure.
3. The probability of heads in a single toss is 0.5, so $p = 0.5$.
4. The tosses should be independent if there is no cheating!

So we can state that:

The number of heads when a fair coin is tossed three times has a B(3, 0.5) distribution, with a probability function $P(x)$, where

$$P(x) = \binom{3}{x}0.5^x(1 - 0.5)^{3 - x} \quad \text{for} \quad x = 0, 1, 2, 3$$

So, when $x = 0$, $P(0) = \binom{3}{0}0.5^0(1 - 0.5)^{3 - 0} = \frac{3!}{0!3!} \times 1 \times 0.5^3 = 0.125$. Similarly, $P(1) = 0.375$, $P(2) = 0.375$, $P(3) = 0.125$.

As a check, we note that the sum of the four probabilities is 1, as it should be since we have a set of four mutually exclusive and exhaustive events.

This distribution can be set out in the form of a table:

Table 6.3 Probability Distribution for the Number of Heads in Three Tosses of a Coin, i.e., the 'B(3, 0.5)'

Number of heads	0	1	2	3
Probability	0.125	0.375	0.375	0.125

6.5 Binomial Probabilities Using Tables and Minitab for Windows

When the number of trials, n, is greater than 3, the calculation of probabilities becomes tedious. Alternative methods are

1. Tables (see Table C.1 in Appendix C), which can be used for certain values of n and p only. This table gives cumulative probabilities, namely the probability of so many or fewer successes.
2. Using Minitab for Windows, which can also be made to list cumulative probabilities, as in Table C.1, for all possible values of n and p, as well as the probability of exactly x successes for all possible values of x.

The use of both methods will be illustrated by the following example.

Example

The B(10, 0.5) distribution, an example of which could be the number of hospital patients, of a random sample of 10 patients, whose illnesses were likely to be cured by a drug known to have a 50% chance of success to date. We should consult the four conditions listed in Section 6.4 in order to check whether the binomial distribution is a good probability model for this example:

1. $n = 10$? Yes.
2. Success = cure, failure = not cured? This might be an over-simplification; there may be partial cures, or cure could be, say, survival for 5 years after the treatment. We really need more information.
3. $p = 0.5$ is given. This is fine if the 50% refers to a defined population of patients and the samples of 10 are truly random (more about samples and populations in Chapter 8).
4. This should be fine unless it is clear that the illness is highly contagious, or occurs in geographical clusters. Again we need to know more in order to check for independence.

Using Table C.1, find a column of numbers for $p = 0.5$ and $n = 10$. Notice that to the right of $n = 10$ there is a column of values for r, and that at the beginning of Table C.1 it states that the probabilities listed are for **r or fewer successes**.

So we can state that, for the B(10, 0.5) distribution, the probability of two or fewer successes, for example, is equal to 0.0547. We can also obtain the probabilities of an exact number of successes by using the result that:

$$P(\text{exactly } r) = P(r \text{ or fewer}) - P((r - 1) \text{ or fewer}),$$

and this is true for $r = 0, 1, 2, \dots , 10$, in this example. It follows from the above and Table C.1 that when $r = 0$, $P(0) = 0.0010 - 0 = 0.0010$, noting that $P(-1) = 0$ since you can't have a negative number of successes.

Similarly for $r = 1, 2, \dots , 10$, we can find $P(1), P(2), \dots , P(10)$. These 11 probabilities are listed in Table 6.4, and of course sum to 1. In order to get Minitab for Windows to produce the two kinds of probability (namely the cumulative and the exact) for the B(10, 0.5) distribution, we must specify the list of possible values for the number of successes in, say, column C1 of the data window, namely the numbers 0, 1, 2, ..., 10. Then:

Choose **Calc > Probability Distributions > Binomial**
Choose **Probability** (in the Binomial distribution window)
Enter **C1** in the **Input Column** box
Enter **10** in the **'number of trials'** box
Enter **0.5** in the **'probability of success'** box
Choose **OK**

The required probabilities should appear in the session window: Table 6.4.

Table 6.4 Binomial Probabilities, from Minitab, for Exact Numbers of Successes When $n = 10$, $p = 0.5$

r	$P(x = r)$
0	0.0010
1	0.0098
2	0.0439
3	0.1172
4	0.2051
5	0.2461
6	0.2051
7	0.1172
8	0.0439
9	0.0098
10	0.0010

In order to get Minitab to produce cumulative binomial probabilities, the procedure is exactly as above, except that we now choose cumulative probability (instead of probability) in the binomial distribution window. Minitab's output includes the information shown in Table 6.5 below:

Table 6.5 Cumulative Binomial Probabilities, from Minitab, When $n = 10$ and $p = 0.5$

r	$P(x$ less or $= r)$
0	0.0010
1	0.0107
2	0.0547
3	0.1719
4	0.3770
5	0.6230
6	0.8281
7	0.9453
8	0.9893
9	0.9990
10	1.0000

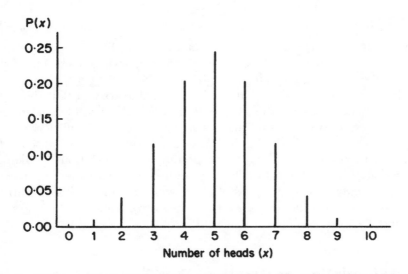

FIGURE 6.1 **Probabilities for a Binomial Distribution with $n = 10$, $p = 0.5$**

Comparing the probabilities produced by using Table C.1 in Appendix C of this text with those produced by Minitab, the agreement is perfect to 4 dps (as it should be). Only you, the student, can decide which method is most suitable for you.

Finally, in this section, we can represent the data in Table 6.4 graphically (see Fig. 6.1). This graph shows that there is no skewness in the distribution. This is because the value of p for this particular binomial distribution is equal to 0.5. Binomial distributions with values of p greater than 0.5 will be negatively skewed, while if p is less than 0.5, the distribution is positively skewed.

In addition, we note that Fig. 6.1 is bell-shaped, a shape we will meet again in Chapter 7 (when we discuss the most famous distribution in statistics!).

6.6 Mean and Standard Deviation of the Binomial Distribution

We quote without proof that the mean and standard deviation of a B(n, p) distribution are

$$\text{mean} = np \qquad \text{standard deviation} = \sqrt{np(1 - p)} \qquad (6.4)$$

Example

What are the mean and standard deviation of the B(5, 0.4) distribution? Since $n = 10$ and $p = 0.5$, mean $= 10 \times 0.5 = 5$, and standard deviation $= \sqrt{10 \times 0.5 \times 0.5} = \sqrt{2.5} = 1.58$.

I hope that it will seem intuitively reasonable to you that the mean number of successes for this distribution should be 5. But what does a standard deviation of 1.58 tell us? As stated before, we will derive more meaning from the value of a standard deviation when we discuss the normal distribution in Chapter 7. For the time being we should remember that the larger the standard deviation, the more the variable will vary. So a variable with a B(10, 0.5) distribution will vary more than a variable with a B(5, 0.4) distribution, since their respective standard deviations are 1.58 and 1.12.

6.7 Simulation of Binomial Distributions Using Minitab for Windows

It is possible to simulate binomial distributions using Minitab for Windows. For example, the B(10, 0.5) distribution is a suitable model for the (binomial experiment of) tossing of a fair coin 10 times and counting the number of heads. Suppose we wish to simulate the repetition of this experiment 100 times. We can do this using Minitab as follows:

Enter 0, 1, 2,…,10 in the first 11 rows of column C1
(these being the possible values of the variable number of
heads in 10 tosses of a fair coin)
Choose **Calc > Random data > Binomial**
Click on **OK**
Enter **100** in the **Generate** box
Enter **C2** in the **Store in columns** box
Enter **10** in the **Number of trials** box
Enter **0.5** in the **Probability of success** box
Click on **OK**
Choose **File > Display data**
Enter **C2** in the box labelled **Columns and constants to display**
Click on **OK**

You should now see a screen display of 100 values of the variable. Then:

Choose **Stat > Basic statistics > Descriptive statistics**
(click on the last of these)
Enter **C2** in the **Variable box**
Click on **OK**

See how close Minitab's MEAN and STDEV are to their theoretical values of 5 and 1.58, respectively.

6.8 Poisson Distribution, an Introduction

The second standard discrete probability distribution we will consider, the Poisson distribution, is concerned with the variable **number of random events per unit time or space**. The word 'random' in this context implies that there is a constant probability that the event will occur in each single unit of time or space (space can be one-, two-, or three-dimensional).

6.9 Some Examples of Poisson Variables

There are many examples to illustrate the great variety of applications of the Poisson distribution as a model for random events.

At the telephone switchboard in a large office block, there may be a constant probability that a telephone call will be received in a given minute. The number of calls received per minute will therefore have a Poisson distribution.

In spinning wool into a ball from the raw state, there may be a constant probability that the spinner will have to stop to untangle a knot. The number of stops per 100 metres of finished wool will then have a Poisson distribution.

In the production of polythene sheeting there may be a constant probability of a blemish (called a 'fish-eye') which makes the film unsightly or opaque. The number of blemishes per square metre will then have a Poisson distribution.

Other examples concerning random events in time are the number of postilions killed by lightning in the days of horse-drawn carriages; the number of major earthquakes in a given country per year; the number of alpha particles emitted per unit time from a radioactive source; and the number of cases of childhood leukaemia per 100,000 children per year.

6.10 The General Poisson Distribution

There is often some confusion between the binomial and the Poisson distribution (in the minds of students!) when they are trying to decide whether a particular variable has either a binomial, a Poisson, or some other distribution. In order to conclude that a variable has a Poisson distribution, we must be able to answer 'yes' to the following questions:

Are we interested in random events per unit time or space?
Is the number of events which might occur in a given unit of time or space theoretically unlimited?

If the answer to the first question is 'yes', but the answer to the second question is 'no', the distribution may be binomial — check the four conditions in Section 6.3.

To calculate Poisson probabilities for a particular variable, we need to know the numerical value of the parameter m for the Poisson. Then we can use either Formula (6.5) for the Probability Function, $P(x)$, of the general Poisson variable, or Table C.2 in Appendix C, or Minitab for Windows (see Section 6.13).

$$P(x) = \frac{e^{-m}m^x}{x!} \quad \text{for} \quad x = 0, 1, 2, \ldots \qquad (6.5)$$

Here $P(x)$ means the probability that x random events will occur per unit time or space; e is the number 2.718... (refer to Section 2.5 if necessary), m is the mean number of random events per unit time or space; and $x = 0, 1, 2, \ldots$ means that we can use Formula (6.5) for $x = 0$ or any positive whole number.

6.11 Calculating Poisson Probabilities, an Example

Example

Suppose that telephone calls arrive randomly at a switchboard at an average rate of 1 call per minute. What are the probabilities that 0, 1, 2, ... calls will be received in a given period of 2 minutes?

Since the probabilities of interest relate to a unit of time of 2 minutes, we must calculate the parameter m as the mean number of calls per 2 minutes.

So, $m = 2$ for this example, and

$$P(x) = \frac{e^{-2}2^x}{x!} \quad x = 0, 1, \ldots$$

Thus

$$P(3) = \frac{e^{-2}2^3}{3!} = 0.180.$$

Substituting other values of x, we obtain Table 6.6.

Table 6.6 Probabilities for a Poisson Distribution for $m = 2$

Number of calls received in 2 minutes (x)	0	1	2	3	4	5
Probability $P(x)$	0.135	0.271	0.271	0.180	0.090	0.036

The probabilities in Table 6.6 do not sum to 1. Why is this? The answer is that x is not restricted to a maximum of 5. However, as you can see from Table 6.6, probabilities for $x > 5$ are very small, in fact $P(x > 5) = 0.017$.

6.12 The Mean and Standard Deviation of the Poisson Distribution

The mean of the Poisson distribution is m as already stated in Section 6.10, and the standard deviation of the Poisson is $\sqrt{m}$. (On the Formula sheet in Appendix A, the result stated in the previous sentence is called Formula [6.6]).

Example

For the example of Section 6.11, $m = 2$, so the mean is 2 and the standard deviation is $\sqrt{2} = 1.41$. Note also that the variance, being the square of the standard deviation (refer to Section 4.12), is also m. So that for any Poisson distribution, mean = variance, both being equal to the parameter m. This property is sometimes used to decide whether a variable has a Poisson distribution, but it is not a very reliable method (see Chapter 16, including Question 10 of Worksheet 16).

6.13 Poisson Probabilities Using Tables and Minitab for Windows

To save time in calculating Poisson probabilities, Table C.2 of Appendix C may be used for certain values of m, instead of the formula method. Table C.2 gives cumulative probabilities, that is, the probabilities of so many or fewer random events per unit time or space.

Example (not the same as that in Section 6.11)

$m = 5$. In Table C.2, find the column of cumulative probabilities for this value of m.

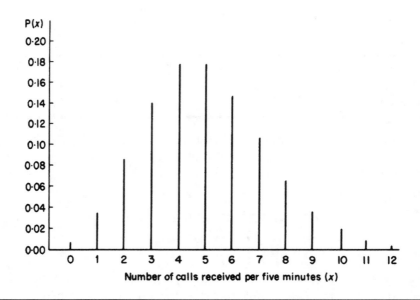

Figure 6.2 Probabilities for a Poisson Distribution with _m_ = 5

To find the probability of, say, at most 10 random events, find the row labelled $r = 10$ and read that:

$$P(10 \text{ or fewer random events when } m = 5) = 0.9863.$$

As with the binomial in Section 6.5, we can obtain the probabilities of an exact number of events by using the result that:

$$p(\text{exactly } r \text{ random events}) = P(r \text{ or fewer random events})$$
$$-P((r - 1) \text{ or fewer random events}),$$

and this is true for $r = 0, 1, 2, \ldots \infty$. Table 6.7 lists the probabilities for $r = 0$ to $r = 10$.

In order to get Minitab for Windows to produce probabilities for the Poisson distribution with a mean $m = 5$ we must enter a list of all possible values for the number of random events per unit time. Although this number is theoretically infinitely large, practically, $0, 1, \ldots, 10$ will be sufficient. How do we know this? The short answer is that we know this already from earlier in this section, and from Fig. 6.2.

Returning to Minitab: enter the numbers $0, 1, 2, \ldots, 10$ into C1. Then:
Choose **Calc > Probability Distributions > Poisson**
Choose **Probability**
Enter **5** in the **Mean** box
Enter **C1** in the **Input column** box
Click on **OK**

The required probabilities should appear in the Session window, as in Table 6.7:

Table 6.7   Poisson Probabilities, from Minitab, for Exact Numbers of Random Events When $m = 5$

r	$P(x = r)$
0	0.0067
1	0.0337
2	0.0842
3	0.1404
4	0.1755
5	0.1755
6	0.1452
7	0.1044
8	0.0653
9	0.0363
10	0.0181

In order to get Minitab to produce cumulative Poisson probabilities, the procedure is exactly as above, except that we now choose Cumulative Probability, instead of Probability in the Poisson Distribution window. Some of the output is as in Table 6.8 below:

Table 6.8   Cumulative Poisson Probabilities, from Minitab, When $m = 5$

r	$P(x \text{ less} = r)$
0	0.0067
1	0.0404
2	0.1247
3	0.2650
4	0.4405
5	0.6160
6	0.7622
7	0.8666
8	0.9319
9	0.9682
10	0.9863

Comparing the probabilities produced by using Table C.2 in Appendix C of this text with those produced by Minitab, the agreement is once again perfect, as expected.

6.14 Simulation of Poisson Distribution Using Minitab for Windows

It is possible to simulate Poisson distributions using Minitab for Windows. For example, suppose we wish to simulate a Poisson with a mean $m = 5$. We can think of this as equivalent to taking a random sample of 100 periods of 5 minutes at a switchboard, assuming we know the number of calls received has a mean of 1 per minute. For this simulation with Minitab:

Enter 0, 1, 2,…,10 in the first 11 rows of column C1 (these being, for all practical purposes, the only values we need to consider)
Choose **Calc** > **Random Data** > **Poisson**
Click on **OK**
Enter **100** in the **Generate** box
Enter **C2** in the **Store in columns** box
Enter **5** in the **Mean** box
Click on **OK**
Choose **File** > **Display Data**
Enter **C2** in the labelled **Columns and constants to display** box
You should now see a screen display of 100 values of the variable.
Choose **Stat** > **Basic Statistics** > **Descriptive Statistics**.
Enter **C2** in the **Variable** box
Click on **OK**
Now compare Minitab's MEAN and STDEV from the simulation with the theoretical values of 5 and 2.236, respectively.

6.15 Poisson Approximation to the Binomial Distribution

There are examples of binomial distributions for which the calculation of approximate probabilities is made easier by the use of the formula or tables for the Poisson distribution! Such an approach can be justified theoretically in the case of binomial distributions having large values of n and small values of p. The resulting probabilities are only approximate, but quite good approximations may be obtained when $p < 0.1$, even if n is not large, by putting $m = np$.

Example

Assume that 1% of people are colour-blind. What is the probability that 10 or more of a random sample of 500 people will be colour-blind?

This is a binomial problem with $n = 500$ and $p = 0.01$. However, Table C.1 (in Appendix C) cannot be used for $n = 500$ and in order to use the binomial Formula (6.3), we would need to calculate $1 - P(0) - P(1) - P(2) - \ldots - P(9)$, quite a tedious calculation. Instead we will use what is called the Poisson approximation to the binomial, for $m = np = 500 \times 0.01 = 5$. Now, from Table C.2, for $m = 5$ and $r = 9$ we read that $P(9$ or fewer colour-blind in a sample of 500 people$) = 0.9682$. So, $P(10$ or more colour-blind in a sample of 500 people$) = 1 - 0.9682 = 0.0318$.

An alternative approach to this example would simply be to use Minitab to generate binomial probabilities as we did in Section 6.7. However, it is, I feel, important to see connections between distributions if they can be demonstrated easily. Otherwise, there is a danger that each distribution may be seen by the student as a different 'rabbit' pulled out of a hat. It is left as an exercise for the reader to check that Minitab gives an answer of 0.0311 to compare with 0.0318 above.

6.16   Summary

The binomial and the Poisson distributions are two of the most important discrete probability distributions. The binomial distribution gives the probabilities for the numbers of successes in a number of (Bernoulli) trials, if four conditions hold. Binomial probabilities may be obtained using Formula (6.3) or, in certain cases, Table C.1 or Minitab for Windows.

The Poisson distribution gives the probabilities for the number of random events per unit time or space. Poisson probabilities may be calculated using Formula (6.5) or, in certain cases, Table C.2 or Minitab for Windows.

If $p < 0.1$, it may be preferable to calculate binomial probabilities using the Poisson approximation to the binomial.

Worksheet 6: The Bernoulli, Binomial, and Poisson Distributions

1. What is a Bernoulli trial?
2. What does the parameter p in a Bernoulli trial stand for?
3. The two outcomes of a Bernoulli trial are usually called success and failure. Which outcome shall I call success, and which failure?

4. What is the general name for a variable which has a binomial distribution?
5. How can you tell *a priori* whether a discrete random variable has a binomial distribution?
6. Why can we think of a Bernoulli distribution as a special case of a binomial distribution?
7. For the distribution B(3, 0.5):
   (a) How many outcomes are there to each trial?
   (b) How many trials are there?
   (c) How many possible values can the variable take?
   (d) What is the mean and what is the standard deviation of this distribution?
   (e) Is this distribution symmetrical? Give a reason for your answer.
Note: questions 8, 9, and 10 are multiple choice. Choose one of the three options in each case.
8. For a binomial distribution with $n = 10$, $p = 0.5$, the probability of 5 or more successes is (a) 0.5, (b) 0.623, (c) 0.377.
9. In a binomial experiment with three trials, the variable can take one of (a) 4 values, (b) 3 values, (c) 2 values.
10. For a binomial distribution with $n = 20$, $p = 0.25$, the probability of 3 or fewer successes is (a) 0.2252, (b) 0.9087, (c) 0.0913.
11. For families with four children, what are the separate probabilities that a randomly selected family will have 0, 1, 2, 3, or 4 boys, assuming that boys and girls are equally likely at each birth? Check that the probabilities sum to 1. Why do they?
    Given 200 families each with four children, how many families would you expect to have 0, 1, 2, 3, or 4 boys?
12. In a multiple-choice test, there are five possible answers to each of 20 questions. If a candidate guesses the answer to each question:
    (a) What is the mean number of correct answers you would expect the candidate to obtain?
    (b) What is the probability that the candidate will pass the test by getting 8 or more correct answers?
    (c) What is the probability that the candidate will get at least one answer correct?
13. In a large batch of items, 5% are defective. If 50 items are selected at random from the batch, what is the probability that:
    (a) At least one will be defective?
    (b) Exactly two will be defective?
    (c) Ten or more will be defective?
    Use tables to answer these questions initially, but check the answers to parts (a) and (b) using a formula and Minitab for Windows.
14. In an experiment with rats, each rat goes into a T-maze in which there is a series of T-junctions. At each junction a rat can turn left

or right. Assuming that a rat chooses at random, what are the separate probabilities that it will make 0, 1, 2, 3, 4, or 5 right turns out of 5 junctions?

15. A new method of treating a disease is estimated to have a 70% chance of effecting a cure. Show that, if a random sample of 10 patients suffering from the disease are treated by this method, the chance that there will be 7 or more cures is about 0.65. Check this answer. What other word could be used instead of 'chance'?

16. Exactly 50 g of yellow wallflower seeds are thoroughly mixed with 200 g of red wallflower seeds. The seeds are then bedded out in rows of 20. Assuming 100% germination,
    (a) Why should the number of yellow wallflower plants per row have a binomial distribution?
    (b) What are the values of n and p for this distribution?
    (c) What is the probability of getting a row with:
       (i) No yellow wallflower plants in it?
       (ii) One or more yellow wallflower plants in it?

17. A supermarket stocks eggs in boxes of six, and 10% of the eggs are found to be cracked. Assuming that the cracked eggs are distributed at random, what is the probability that a customer will find that a box he chooses contains: (a) no cracked eggs? (b) at least one cracked egg? If he examines five boxes, what is the probability that three or more will contain no cracked eggs?

18. For the Poisson distribution, we use Formula (6.5):

$$P(x) = \frac{e^{-m}m^x}{x!} \quad \text{for} \quad x = 0, 1, 2, \ldots$$

What do the symbols m, e, and x stand for?
What values can x take?

19. If a Poisson distribution variable has a mean of 4, what is its standard deviation and what is its variance? What can you say about the mean and variance of any Poisson distribution?

20. The Poisson distribution is the distribution of the number of random events per unit time. What does the word 'random' mean here?

Note: questions 21 and 22 are multiple choice. Choose one of the three options in each case.

21. For a Poisson distribution with a mean $m = 2$, $P(2)$ is equal to (a) 0.2707, (b) 0.5940, (c) 0.7293.

22. For a Poisson distribution with a mean 10 per unit time, the probability of at least 23 random events per unit time is (a) 0.0003, (b) 0.0002, (c) 0.0004.

23. Assuming that breakdowns in a certain electricity supply occur randomly with a mean of one breakdown every 10 weeks, calculate the separate probabilities of 0, 1, and 2 breakdowns in any period of 1 week.

24. Assume that the number of misprints per page of a book has a Poisson distribution with a mean of one misprint per five pages. What percentage of pages contain no misprints? How many pages would you expect to have no misprints in a 500-page book?

25. A hire firm has three ladders which it hires out by the day. Records show that the mean demand is 2.5 ladders per day. If it is assumed that the demand for ladders follows a Poisson distribution, what is
    (a) The percentage of days on which no ladder is hired?
    (b) The percentage of days on which all three ladders are hired?
    (c) The percentage of days on which demand outstrips supply?

26. A roll of cloth contains an average of three defects per 100 square metres distributed at random. What is the probability that a randomly chosen section of 100 square metres of cloth contains:
    (a) No defects?
    (b) Exactly three defects?
    (c) Three or more defects?

27. A rare blood group occurs in only 1% of the population, distributed at random. What is the probability that at least one person in a random sample of 100 has blood of this group? Use both the binomial method and the Poisson approximation to the binomial method. Compare your answers. Which is correct?

28. If, in a given country, an average of 1 miner in 2000 loses his life due to accident per year, calculate the probability that a mine in which there are 8000 miners will be free from accidents in a given year.

29. The average number of defectives in batches of 50 is 5. Obtain the probability that a batch will contain:
    (a) 10 or more defectives.
    (b) Exactly 5 defectives.
    Use both the binomial and the Poisson approximation to the binomial methods and compare your answers.

30. Geometric Distribution (optional question, see below).

Background

Another discrete probability distribution which can be useful is the **geometric distribution**. At first it may seem like the binomial, because it also concerns Bernoulli trials. However there are two principal differences, which are linked together:

1. In the binomial the number of trials is fixed; in the geometric the number of trials varies.
2. In the binomial the number of successes is the variable; in the geometric the number of successes is fixed (and always equals 1).

The variable in the case of the **geometric distribution**, in general terms, is

the number of Bernoulli trials up to and including the first success

(recall that in a Bernoulli trial there are only two possible outcomes, which we call success and failure, and p, the probability of success, is constant). *Note: I think of the geometric as 'the Driving-Test distribution':* *you have a number of failures, i.e., 0, 1, 2,..., followed by one success* *(and then you stop taking the test).*

It is easy to obtain the probability function for the geometric. Let x stand for the variable number of trials up to and including the first success. Then there must have been $(x - 1)$ failures prior to the one success. Since the trials are independent, we can use the special case of the multiplication law (for statistically independent events) to show that:

$$P(x) = P(failure) \times P(failure) \times \ldots \times P(success)$$
$$= (1 - p)(1 - p)\ldots p$$

Since the trials are stopped after the first success, the smallest number of trials must be 1. Hence:

$$P(x) = (1 - p)^{x-1}p \quad \text{for} \quad x = 1, 2,\ldots$$

Now for the Question: Can you show that the probabilities for the geometric distribution sum to 1? You may quote the result that:

$$1 + x + x^2 + x^3 + \ldots = \frac{1}{1 - x} \quad \text{if } x \text{ is positive and less than } 1.$$

Chapter 7

Continuous Probability Distributions

7.1   Introduction

In Chapter 3 we considered an example of a continuous variable, namely, the height of students, and we summarised the heights of 27 female students by drawing a histogram (see Worksheet 5, Question 16), reproduced here as Fig. 7.1. We also saw in that question how to express a probability as the ratio of two areas, so that we could make statements such as:

P (randomly selected female student has a height between 164.5 and 169.5 cm) $= \dfrac{\text{Area of rectangle on base } 164.5 - 169.5}{\text{Total area of histogram}}$

Suppose we apply this idea to the heights of all female students in their first year at a university in the U.K. In the histogram the number of students in each group would now be much greater, so we could afford to have many more than six groups, and still have a fairly large number of students in each group. The histogram would look something like Fig. 7.2, where the vertical sides of the rectangles have been omitted and the tops of the rectangles have been smoothed into a curve.

If this graph is 'scaled' in the vertical direction so that the total area under the curve is 1 in some units, then we would be wrong to keep calling the vertical axis 'Number of students'. However, this curve would

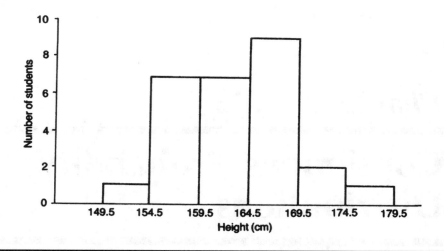

Figure 7.1   Histogram of the Heights of 27 Female Students

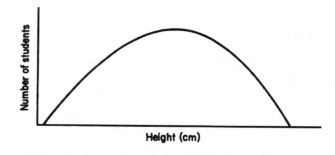

Figure 7.2   Histogram of the Heights of all Female Students in Higher Education

have the property that the probability of a female student's height being between any two values would be equal to the area under the curve between these values, as shown in Fig. 7.3. For example,

P(randomly selected female student has a height between 164.5 and 169.5 cm) = Area under curve between 164.5 and 169.5 cm.

Assuming such a curve can be drawn, it is an example of the graphical representation of a **continuous probability distribution**. Compare Fig. 7.3 with Fig. 6.1, an example of the graphical representation of a **discrete** probability distribution.

There are several standard types of continuous probability distribution. We will consider two of the most important, namely, the normal distribution and the rectangular (which Minitab refers to as the 'continuous uniform' distribution).

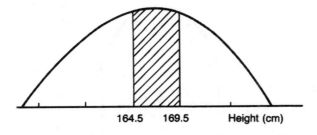

164.5   169.5   Height (cm)

Figure 7.3 Continuous Probability Distribution for the Variable 'Height'

7.2 The Normal Distribution

The normal distribution is the most important in Statistics. There are two main reasons for this:

1. It arises when a variable is measured for a large number of nominally identical objects, and where the variation may be assumed to be caused by a number of factors, each exerting a small positive or negative random influence on an individual object. An example is the variable 'height of a female student', where the variation in heights is caused by many factors such as age, diet, exercise, heights of parents, bone structure, and so on.
2. The properties of the normal distribution have a very important application in the theory of Statistical Inference, which is what statisticians call 'drawing conclusions from sample data about the larger population from which the sample was drawn'. The methods which are based on this theory will be discussed in every chapter from now on, including situations in which the variable of interest is demonstrably not normally distributed!

Returning to the idea of graphically representing distributions, the normal distribution has a **symmetrical bell shape**, with most values concentrated towards the middle, a few extreme values, and it is **unimodal** (i.e., one peak, see Fig. 7.4). It has two **parameters**, μ and σ.

At this point you should note the following important observations regarding notation. In Chapter 4, the symbols $\bar{x}$ and s were used to denote the **sample** mean and **sample** standard deviation, respectively. The Greek symbols μ (lower-case mu) and σ (lower-case sigma) are used here because we are now dealing with a **population** of measurements. Samples and populations will be defined and discussed more fully in Chapter 8.

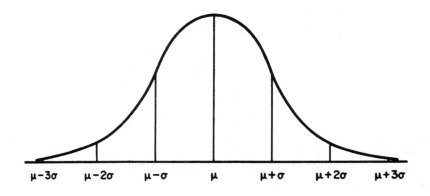

$\mu-3\sigma$ $\mu-2\sigma$ $\mu-\sigma$ μ $\mu+\sigma$ $\mu+2\sigma$ $\mu+3\sigma$

Figure 7.4   The Normal Distribution: μ is the Mean of the Distribution, and σ is the Standard Deviation of the Distribution

There are a number of related properties of the normal distribution which (at last!) give us a better understanding of the meaning of standard deviation as a measure of variation:

1. Approximately 68% of the area under any normal distribution curve lies within one standard deviation of the mean. So the area between the vertical lines drawn at $(\mu - \sigma)$ and $(\mu + \sigma)$ in Fig. 7.4 is roughly two thirds of the total area. Recall that the total area is equal to 1.
2. Approximately 95% of the area under any normal distribution curve lies within two standard deviations of the mean. To be more precise, we can quote that exactly 95% of the area lies within 1.96 standard deviations of the mean.
3. Approximately 99.7% of the area under any normal distribution curve lies within three standard deviations of the mean.

7.3   An Example of a Normal Distribution

Suppose that we know that the variable 'height' (of all female students in their first year in a U.K. university) is normally distributed with a mean $\mu = 163$ cm and standard deviation $\sigma = 6$ cm.

Using the properties stated in the previous section we could state, for example, that approximately 95% have heights between $163 - 2 \times 6 = 151$ cm and $163 + 2 \times 6 = 175$ cm. This is equivalent to the statement that 'the probability that a randomly selected female student will have a height between 151 and 175 cm is 0.95'.

But how can we calculate probabilities and percentages for other heights of interest? The answer is that we need to be able to obtain areas under any normal distribution curve. One way of doing this is to use Table C.3(a) in Appendix C. (Another way is to use Minitab, as we shall see in Section 7.4.) Table C.3(a) enables us to calculate probabilities for any normal distribution if we know numerical values for μ and σ. The table actually gives probabilities in terms of areas of the normal distribution curve; namely, areas to the left of particular values of the variable.

Consider the example of the normal distribution of heights given above, i.e., with a mean $\mu = 163$ cm and $\sigma = 6$ cm. In shorthand form we may refer to this as the $N(163, 6^2)$ distribution, where the **general normal distribution is** $N(\mu, \sigma^2)$. Note that the following five questions in this section refer to this particular example.

Question 7.1

What is the probability that a randomly selected female student has a height greater than 170 cm?

The answer is the area to the right of 170 in Fig. 7.5, since 'to the right of 170' implies 'greater than 170'.

In order to use Table C.3(a), we first have to 'transform' our normal distribution into one with a mean $\mu = 0$ and standard deviation $\sigma = 1$, the so-called **standardized normal distribution**. We do this by calculating 'z values' using the formula

$$z = \frac{(x - \mu)}{\sigma} \tag{7.1}$$

Let's see how to apply this to our example.

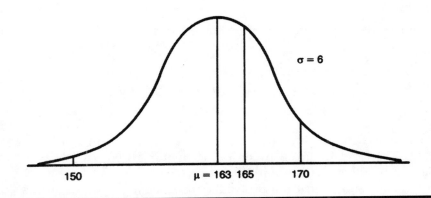

Figure 7.5   A Normal Distribution With $\mu = 163$, $\sigma = 6$

Since we are interested in the value 170 cm in this question, let $x = 170$ for the moment. Now we calculate the z value, using $\mu = 163$ and $\sigma = 6$. Hence:

$$z = \frac{(170 - 163)}{6} = 1.17$$

Using Table C.3 (a) for $z = 1.17$, we read that the area to the left of 170 cm is 0.8790. Since the total area under the curve is 1, the area to the right of 170 is $1 - 0.8790 = 0.1210$. We can also state that 12.1% (about 1 in 8) of female students have a height greater than 170 cm, using the idea that percentage $=$ probability $\times$ 100 (see Section 5.6).

One useful way of understanding what a 'z value' means is to think of it as 'the number of standard deviations we are from the mean of the distribution'. It is also useful to realise that values of the variable greater than the mean give rise to positive values of z, while negative values of z arise if the value of the variable is less than the mean. It should come as no surprise that $z = 0$ when the value of the variable is equal to the mean!

Question 7.2

What is the probability that height lies between 165 and 170 cm for the same distribution as above?

The answer is the area between 165 and 170 in Fig 7.5, which we can think of as:

area to the left of 170 $-$ area to the left of 165

Since we are now interested in the value 165, let $x = 165$. Now calculate the z value:

$$z = \frac{(165 - 163)}{6} = 0.33$$

Using Table C.3(a) for $z = 0.33$, we read that the area to the left of 165 is 0.6293. Therefore the area between 165 and 170 $= 0.8790 - 0.6293 = 0.2497$, using the answer to Question 7.1.

We have shown that the probability that height lies between 165 and 170 is 0.2497. We can also state that 24.97% of female students (about 1 in 4) have a height between 165 and 170 cm.

Question 7.3

What is the probability that height is less than 150 cm?
 The answer is the area to the left of 150 in Fig. 7.5. Let $x = 150$, then

$$z = \frac{(150 - 163)}{6} = -2.17$$

The negative sign for z indicates that the value of x is less than the mean, which we can also see from Fig. 7.5. The area given in the Table for $z = +2.17$ is 0.9850. Hence the area to the right of $z = 2.17$ is $1 - 0.9850 = 0.015$. By symmetry, this is also to the left of $z = -2.17$. So the required probability is 0.015 (1.5% or about 1 in 70).

Question 7.4

What is the probability that height lies between 150 and 165 cm?
 From previous answers, the required probability is

$$0.6293 - 0.0150 = 0.6143$$

Question 7.5

What is the probability that height is less than 163 cm?
 By the symmetry of the normal distribution shown in Fig. 7.5, the answer is 0.5, which we can verify by using $x = 163$ and $z = (163 - 163)/6 = 0$, and Table C.3(a).

7.4 Normal Probabilities Using Minitab for Windows

The Minitab for Windows method for Questions 7.1 to 7.5 is as follows:
 Choose **Calc > Probability Distributions > Normal**
 Choose **Cumulative probability**
 Enter **163** in the **Mean** box
 Enter **6** in the **Standard deviation** box.
 Choose **Input constant**
 Enter **170** in the **Input constant** box
 Click on **OK**
The following output should appear on the screen:
 MTB > CDF 170;
 SUBC > Normal 163 6.
 170.0   0.8783

This output indicates that the area to the left of 170 for the N (163, 6^2) distribution is 0.8783, and this is also the required probability. Our answer is slightly different from the 0.8790 we obtained using Table C.3(a), because we rounded our z value to 2 dps.

CDF stands for cumulative distribution function, which is not of any great interest to us except for the fact that there is an Inverse CDF, for when we want to answer a question the 'other way round'. For example, we could ask:

'What height is such that 90% of students have a height less than it?' We can answer this using the same method as above except:

(i) We choose **Inverse cumulative probability**, instead of Cumulative probability,

(ii) We enter **90**, instead of 170, in the **Input constant** box.

The output should be
MTB > InvCDF 90;
SUBC > Normal 163 6.
   0.90   170.7

Question 7.2: enter **165** as the input constant, and use the answer to Question 7.1.

Question 7.3: enter **150** as the input constant.

Question 7.4: this can be answered from answers to earlier questions.

Question 7.5: quote symmetry, or enter **163** as the **input constant**.

7.5 Simulation of the Normal Distribution Using Minitab for Windows

It is possible to make Minitab simulate values from a specified normal distribution:

Choose **Calc** > **Random data** > **Normal**
Enter **100** in **Generate** box
Enter **C1** in **Store in column** box
Enter **163** in **Mean** box
Enter **6** in **Standard deviation** box.
Click on **OK**
Choose **Print C1**.

There should now be a random sample of 100 values from the $N(163, 6^2)$ distribution in C1. We can print the values in a list using File and Print Window, or we can summarise the 100 values using:

Stat > Basic Stats > Descriptive Stats,
and then enter **C1** in the **Variables** box, and finally click on **OK**. The screen output should be like Table 4.1, and the values for the mean, median, and standard deviation, for example, should be close to the equivalent theroretical values for the $N(163, 6^2)$ distribution, namely, 163, 163, and 6, respectively. If we repeat the simulation, this time for a much larger sample of 10,000, say, we would expect to get even closer to the theoretical values.

One purpose of simulation is that we can take samples from a known population by sitting at a desk. More importantly, we can use simulation to make some aspects of statistical theory believable, without having to use advanced mathematics to 'prove' this theory rigorously.

7.6   Rectangular Distribution

Because of the great importance of the normal distribution, students who have taken introductory courses in statistics tend to believe that all continuous variables are normally distributed. Partly to counteract this erroneous belief at an early stage, we now introduce another continuous probability distribution, namely, the **rectangular** (also called the **continuous uniform**) distribution. The erroneous belief will also be counteracted in Chapter 11 when we deal with inferential methods which deal specifically with nonnormal continuous variables. The rectangular is a rather dull and flat distribution (Fig. 7.6), but it does have the advantage that probabilities are easy to calculate.

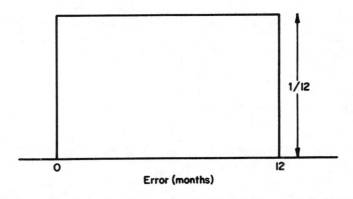

Figure 7.6   A Rectangular Distribution for the 'Error' in a Stated Age

Example

Suppose we consider the 'error' which is made when a person states his or her 'age at last birthday.' The error is the difference:

$$\text{actual age} - \text{age last birthday}$$

and this continuous variable is equally likely to lie anywhere in the range 0 to 12 months, so that its probability distribution is as in Fig. 7.6. Note that, since the total area of the rectangle must be equal to 1, the height of the rectangle must be 1/(base) = 1/12.

Question 7.6

What percentage of errors will be less than 3 months?

The probability of an error of less than 3 months is the area to the left of 3 in Fig. 7.6, which is 3 × 1/12 = 1/4, so 25% of errors will be less than 3 months.

7.7 The Normal Approximation to the Binomial Distribution

Just as there are conditions (see Section 6.15) when the calculation of approximate binomial probabilities is made easier by using the formula or tables for the Poisson distribution, so there are conditions when it is preferable to use normal distribution tables to obtain approximate binomial probabilities.

We may use the so-called 'normal approximation to the binomial' when $np > 5$, and $n(1 - p) > 5$. These conditions are more likely to be met if n is large and p is not too close to 0 or 1.

Example

Suppose that one person in six is left-handed. If a class contains 40 students, what is the probability that 10 or more will be left-handed? Assuming that the four conditions for the binomial apply, this is a binomial problem with $n = 40$, $p = 1/6$, so $np = 6.67$ and $n(1 - p) = 33.33$, and hence the conditions for for using the 'normal approximation to the binomial' are satisfied. We may therefore treat the variable 'number of left-handed students in a sample of 40' as though it was normally

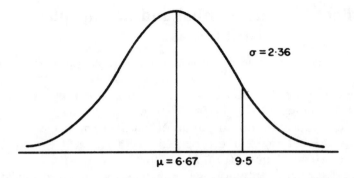

Figure 7.7   A Normal Distribution With $\mu = 6.67$, $\sigma = 2.36$

distributed with:

$$\mu = np = 6.67 \quad \text{and} \quad \sigma = \sqrt{np(1 - p)} = 2.36$$

The distribution is shown in Fig. 7.7. Before we use Table C.3(a), we should apply a continuity correction of 0.5 since the 'number of left-handed students' is a discrete variable while the normal distribution is continuous. Since '10 or more on a discrete scale' is equivalent to 'more than 9.5 on a continuous scale,' we use $x = 9.5$ to obtain the value of z:

$$z = \frac{9.5 - 6.67}{2.36} = 1.2$$

giving an area to the left of 9.5 of 0.8849, using the table. Hence the probability of 10 or more left-handed players in 40 is $1 - 0.8849 = 0.1151$, or 11.5%.

7.8   Summary

The normal and the rectangular distributions are two standard types of continuous probability distribution. The normal distribution is the most important in statistics because it arises when a number of factors exert small positive or negative effects on the value of a variable, and because it is extremely useful in the theory of statistical inference.

Probabilities and percentages for the normal distribution may be obtained using tables or Minitab when we have numerical values for μ and σ, and for the rectangular distribution by calculating the areas of rectangles. The total area under any continuous distribution curve is 1.

The normal distribution tables may also be used to obtain approximate binomial probabilities if $np > 5$ and $np(1 - p) > 5$, and where it is not possible to use (binomial) Table C.1.

Worksheet 7: The Normal and Rectangular Distributions

Questions 1 to 4 are multiple choice. Choose one option in each case.

1. In a normal distribution with $\mu = 10$ and $\sigma = 4$, the probability of exceeding 13 is (a) 0.0668, (b) 0.2266, (c) 0.9332.
2. A continuous random variable has a normal distribution with $\mu = 10$ and $\sigma = 2$, the probability of a value of exactly 10 is (a) 0, (b) 0.40 (2 dps), (c) 0.20 (2 dps).
3. In a normal distribution with mean μ and standard deviation σ:
   (a) 10% of the values are outside the range $(\mu - 1.645\sigma)$ to $(\mu + 1.645\sigma)$.
   (b) 10% of the values are greater than $(\mu + 1.645\sigma)$.
   (c) 10% of the values are outside the range $(\mu - 1.96\sigma)$ to $(\mu + 1.96\sigma)$.
4. In order to test the effectiveness of a drug, 12 individuals are treated with it. The variable of interest is the number of people who recover after taking the drug. The appropriate distribution to use in this case is the: (a) Poisson, (b) normal, (c) binomial.
5. For the binomial and Poisson distributions the probabilities sum to 1. What is the equivalent property for the normal and rectangular distributions?
6. The weights of 5p oranges are normally distributed with mean 70 g and standard deviation 3 g. What percentages of these oranges weigh: (a) over 75 g, (b) under 60 g, (c) between 60 and 75 g? Given a random sample of 50 5p oranges, how many of them would you expect to have weights in categories (a), (b), and (c)? Suppose that the oranges are kept for a week and each loses 5 g in weight during this time. Rework this question.
7. A fruit grower grades and prices oranges according to their diameter:

Diameter (cm)	Price per orange (p)
Below 5	4
Above 5 and below 6	5
Above 6 and below 7	6
Above 7 and below 8	7
Above 8 and below 9	8
Above 9 and below 10	9
Above 10	10

Assuming that the diameter is normally distributed with a mean of 7.5 cm and standard deviation 1 cm, what percentages of oranges will be priced at each of the above prices? Given 10,000 oranges: (a) what is their total price, (b) what is their mean price?

8. A machine produces components whose thicknesses are normally distributed with a mean of 0.4 cm and standard deviation 0.02 cm. Components are rejected if they have a thickness outside the range 0.38 to 0.41 cm. What percentage are rejected? Show that the percentage rejected will be reduced to its smallest vaue if the mean is reduced to 0.395 cm, assuming the standard deviation remains the same.

9. Guests at a large hotel stay for a mean of 9 days with a standard deviation of 2.4 days. Among 1000 guests how many can be expected to stay: (a) less than 7 days, (b) more than 14 days, (c) between 7 and 14 days?
Assume that length of stay is normally distributed.

10. It has been found that the annual rainfall in a town has a normal distribution, because of the varying pattern of depressions and anticyclones. If the mean annual rainfall is 65 cm, and in 15% of years the rainfall is more than 85 cm, what is the standard deviation of the annual rainfall? What percentage of years will have a rainfall of less than 50 cm?

11. In the catering industry the wages of a certain grade of part-time kitchen staff are normally distributed with a standard deviation of £12. If 20% of staff earn less than £90 a week, what is the mean wage? What percentage of staff earn more than £125 a week?

12. Sandstone specimens contain varying percentages of void space. The mean percentage is 15%, the standard deviation is 3%. Assuming that the percentage of void space is normally distributed, what proportion of specimens have a void space: (a) less than 15%, (b) less than 20%, (c) less than 25%?

13. The height of adult males is normally distributed with a mean of 172 cm and standard deviation of 8 cm. If 99% of adult males exceed a certain height, what is this height?

14. A machine, which automatically packs potatoes into bags, is known to operate with a mean of 25 kg and a standard deviation of 0.5 kg. Assuming a normal distribution, what percentage of bags weigh: (a) more than 25 kg, (b) between 24 and 26 kg? To what new target mean weight would be exceeded by 0.1% of bags?

15. Referring to Fig. 7.4, let $x = (\mu + \sigma)$. Show that this implies a 'z value' of +1, irrespective of the values of μ and σ. Hence confirm the first of the three numbered statements, labeled number 1, made

at the end of Section 7.2. Now confirm the other two statements labelled 2 and 3.

16. A haulage firm has 60 lorries. The probability that a lorry is available for business on any given day is 0.8. Find the approximate probability that, on any given day:
    (a) 50 or more lorries are available.
    (b) Exactly 50 lorries are available.
    (c) Fewer than 50 lorries are available.
    Now find the exact probabilities using Minitab, for parts (a), (b), and (c).

17. An office worker commuting from Oxford to London each day keeps a record of how late his train arrives in London. He concludes that the train is equally likely to arrive at any time between 10 minutes late and 40 minutes late. If the train is more than 30 minutes late the office worker will be late for work. How often will this happen? Rework this question assuming now that the number of minutes late is normally distributed with $\mu = 25$ mintes, $\sigma = 7.5$.
    Represent both distributions in a sketch, indicating the answers as areas in the sketch.

Chapter 8

Samples and Populations

We should extend our views far beyond the narrow bounds of
a parish, we should include large groups of mankind.

8.1 Introduction

In each of the remaining chapters we shall be concerned mainly with
statistical inference, by which we mean:

**Drawing conclusions from sample data about the larger
populations from which the samples are drawn.**

Although we have met the words 'sample' and 'population' in earlier
chapters, they were not defined. We now present the definitions:

A **population** is the whole set of measurements or counts about which
we want to draw a conclusion.

If we are interested in only one variable, we call the population **univariate**.
For example, the heights of all female students in their first year at university
form a univariate population. Notice that a population is a set of measurements or counts, not the individuals or objects on which the measurements
or counts are made.

A **sample** is a subset of the population, a set of some of the measurements or counts which comprise the population.

You might find that it helps to understand 'sample and population' by means
of a visual image. In the student height example, assume that each qualifying
student (female and first-year) has written her height in centimetres on a
standard piece of plastic card and then dropped the card into a very large sack.

If there are, say, 400,000 qualifying students, the bag should ultimately contain this number of cards. The 400,000 heights are the population. A sample is any subset of this. Suppose that we thoroughly mix the cards and select one without looking, write down the height, and then replace the card in the sack. Repeat this procedure 20 times in all. The 20 heights are a sample from the population.

The procedure used and why we selected 20 in the sample will be discussed in this chapter.

8.2 Reasons for Sampling

The first and most obvious reason for sampling is to save time, money, and effort. For example, in opinion polls before a general election it would not be practicable to ask the opinion of the whole electorate on how each individual intends to vote.

The second and less obvious reason for sampling is that, even though we have only part of all the information about the population, nevertheless the sample data can be useful in drawing conclusions about the population, provided that we use an appropriate **sampling method** (Section 8.3) and choose an appropriate **sample size** (Section 8.4).

The third reason for sampling applies to the special case in which the act of measuring the variable destroys the individual, such as in the destructive testing of explosives. Clearly, testing a whole batch of explosives would be ridiculous.

8.3 Sampling Methods

There are many ways of selecting a sample from a population, but the most important method is **random sampling**. Indeed, the methods of statistical inference used in this book apply only to cases in which the sampling method is random. Nevertheless, there are other sampling methods which are worth mentioning for reasons stated below.

A **random sample** is defined as one for which each measurement or count in the population has the same chance (probability) of being selected. A sample selected so that the probabilties are not the same for each measurement or count is said to be a **biased** sample. For the population of female student heights, selecting only those who were members of netball teams might result in a biased sample since they would tend to be taller than the average female student. Random sampling requires that we can identify all the individuals or objects which comprise the population, and that each measurement or count to be included in the sample is chosen using some method which ensures equal probability, for example, by the use of random number tables such as Table C.4 of Appendix C.

However, perfectly random samples are often difficult to obtain, even if we have a 'complete' list of the population. For example, nonresponse may be a problem; in a house-to-house survey in which the person carrying out the survey needs to speak to the owner-occupier or tenant, it is not unknown for up to 25% of the targeted individuals to be 'not at home', even if repeated calls are made at different times and on different days.

Example of How to Take a Random Sample

For the student height example, suppose that each student is assigned a unique six-digit number (six because 400,000, the size of the population, has six digits). Starting anywhere in Table C.4, read off six consecutive digits by moving from one digit to the next in any direction: up, down, left, right, or diagonally. This procedure could result in the number 0 6 1 9 7 8, say. If there is a student with this number, then her height would be the first in the sample, and the second random number would be selected. If there is no student with this number, ignore this number, and obtain another by the same means and so on. We carry on until the required number of observations, called the **sample size**, has been reached. Random numbers are available on most scientific calculators and computers.

Where the individuals are objects which are fixed in a given location, the method of random sampling can also be used. There are a number of examples in geography, geology, and environmental biology where such samples are required. One method of obtaining a random sample is to overlay a plan of the area to be investigated with a rectangular grid which includes all the points of interest in the area (Fig. 8.1).

From one corner of the rectangle, one side is labelled with the 100 values, 00 to 99, spaced at equal intervals. The other side is labelled in

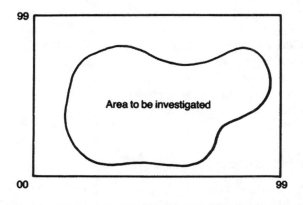

Figure 8.1   Random Sampling from a Two-Dimensional Area

the same way. By selecting two two-digit random numbers, the coordinates of a randomly selected point within the rectangle are determined. If this point falls outside the area of interest it is ignored. The procedure is repeated until the required number of points has been selected.

Systematic sampling may be used to cut down the time taken to select a random sample. In the student height example, suppose that we require a sample of 4000 from a population of 400,000. (This could be described as a '1 in 100' sample since 400,000/4000 = 100, or a 1% sample since 1% means 1 in 100.) We start by selecting one random number only in the range 000,000 to 000,099. Then we derive all other selected student numbers by adding 100, 200, 300, and so on. Suppose that we initally select student number 000,057. Then the next two to be selected are 000,157 and 000,257..., and the last in a sample of 4000 would be 399,957 (assuming there are exactly 400,000 in the population with consecutive numbers 000,000 to 399,999).

Systematic sampling is suitable in that it provides a quasi-random sample so long as there is no periodicity in the population list or geographical arrangement which coincides with the selected numbers. For example, if we were selecting from a population of houses in an American city where the streets are laid out in a grid formation, then selecting every 100th house could result in always choosing a corner house on the intersection of two streets or avenues, possibly giving a biased sample.

Stratified sampling may be used where it is known that the individuals or objects to be sampled comprise not one population but a number of distinct subpopulations or **strata**. These strata may have quite different distributions for the variable of interest.

Example

For the population of the heights of female students in their first year at a U.K. university, we may wish to separate Arts and Science students. Suppose that 65% of these students are taking Arts subjects while the remaining 35% are taking Science subjects. We could take a 1 in 100 sample from each of these two strata to provide an overall 1% sample.

Other methods include quota sampling, cluster sampling, multistage and sequential sampling. Each method has its own special application area and will not be discussed here.

8.4 Sample Size

The most common question asked by an investigator who wishes to collect and analyse data is 'How much data should I collect?' We refer to the number of observations to be included in the sample as the **sample size**, so the investigator should be asking: 'What sample size should I choose?' If we wish to give a slightly more sophisticated answer than, for example, 'a sample size of 20 seems a bit too small, but 30 sounds about right', we can use the arguments in the following example.

Example

Suppose we wish to estimate the (population) mean height, μ, of first-year female undergraduates. The required sample size should depend on two factors:

1. The precision we require for the estimate which the investigator must specify, knowing that the greater precision he or she requires, the larger the required sample size will be.
2. The variability of the variable height, as measured by its standard deviation. We would think that the larger the standard deviation is, the larger the required sample size will be. The standard deviation can only be determined when we have some data. But since we are trying to decide how much data to collect, we are in a chicken-and-egg situation. We might be lucky if we can obtain a rough estimate of the s.d. from previous work we or someone else has carried out, but if there isn't any, then a pilot survey of, say, a sample of 20 or 30 will provide an approximate value for standard deviation.

Another method which can be used if we know that our variable is normally distributed is as follows. For height, say, what height is only exceeded by about 1 in 1000 individuals in our population? For female students you might guess 190 cm. Then ask yourself, what height is such that all except 1 in 1000 students exceed it? Your answer might be 145 cm. Then using the standard result that 99.8% of any normal distribution lies within 3.09 s.d. of the mean (refer to Section 7.2 if necessary) it follows that:

$$2 \times 3.09 \times sd = 190 - 145 = 45.$$

So,

$$sd = \frac{45}{6.18} = 7.3 \text{ cm.}$$

Since this is only a very approximate answer, we might call it 7 cm. We will return to a discussion of sample size when we have discussed the ideas of confidence intervals in the next chapter.

8.5 Sampling Distribution of the Sample Mean

This is not a book about the mathematical theory of statistics, but the theory which we now discuss is essential for a more complete understanding of the inferential methods to be discussed later, particularly those concerning confidence intervals (in Chapter 9).

Suppose that we are sampling from a population of measurements which has a mean, μ, and a standard deviation, σ. Note the use of Greek symbols μ and σ for the mean and standard deviation, since we are talking about the **'parameters of a population'** (see Section 7.2).

We will suppose that the measurement is of a continuous variable, so that its probability distribution is continuous but not necessarily normal, for example, like Fig. 8.2.

Suppose we select a random sample of size n from this 'parent' population and calculate $\bar{x}$, the sample mean, using the methods of Section 4.2). If we continue this procedure of taking samples of size n and calculating the sample mean on each occasion, we will eventually have a large number of sample means(!). Since the random samples are very unlikely to consist of exactly the same observations, these sample means will vary and form a new population with a distribution which is known as the:

sampling distribution of the sample mean

This distribution will have a mean and a standard deviation, which we denote by $\mu_{\bar{x}}$ and $\sigma_{\bar{x}}$ to distinguish them from the μ and σ of the parent population. The sampling distribution will obviously have a shape. The question of interest, which is really three questions in one, is the following: *What is the connection between the mean, standard deviation, and shape of the distribution of the parent population and the mean, standard deviation, and shape of the sampling distribution of the sample mean?*

The answer to this question is supplied now by some of the most important results in statistical theory which will be quoted (without proof)

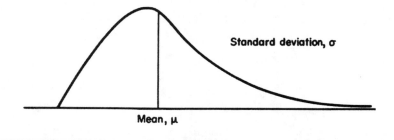

Figure 8.2   A Continuous Probability Distribution

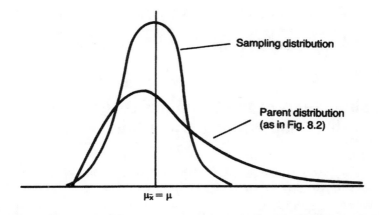

Figure 8.3   A 'Parent' Distribution and the Sampling Distribution of the Sample Mean

in three parts:

1. $\mu_{\bar{x}} = \mu$ (8.1)

2. $\sigma_{\bar{x}} = \dfrac{\sigma}{\sqrt{n}}$ (8.2)

3. If n is large, the distribution of the sample mean is approximately normal, irrespective of the shape of the distribution of the parent population. (If, however, the parent is normal, then the distribution of the sample mean is also normal for all n.)

These three important results can be expressed in words and also graphically (see Fig. 8.3). In words, the three parts of the theory state that if a large number of random samples of size n are taken from a parent population, then:

1. The mean of the sample means will equal the mean of the parent.
2. The standard deviation of the sample means will be smaller than the standard deviation of the parent by a factor of $\sqrt{n}$. For example, if $n = 4$, the standard deviation of the sample means will be half the standard deviation of the parent (since $\sqrt{4} = 2$).
3. The shape of the sampling distribution of the sample mean is approximately normal for large sample sizes, even if the parent distribution is not normal (and the larger the sample size, the more closely is this distribution normal).

The third result is called the **central limit theorem** and is an example of one of the two reasons (see Section 7.2) for the importance of the normal distribution in Statistics.

In the next chapter we will use the three results quoted above to make estimates of the mean of a population from sample data (these estimates are called **confidence intervals**).

8.6 Simulation of the Sampling Distribution of the Sample Mean Using Minitab

The statistical theory in the previous section was quoted without proof because the proof is beyond the scope of this book. However, we can use simulation to show whether the theory seems reasonable by taking samples from a population whose mean, standard deviation, and shape are 'known' to us, and seeing if the parts of the theory are satisfied. This is an artificial and unreal situation because in the 'real world' we have only estimates of the mean and standard deviation of the parent population, based on the sample data we collect. In the real world we regard μ and σ as **unknown population parameters**.

Suppose our parent population is $N(163, 6^2)$ and that we want to take 200 samples each of size 9 from this population. We will store the first nine values in row 1, columns 1 to 9, the next nine in row 2, and so on until we have completed 200 rows. Then the mean of each row is calculated and put in column 10. The 200 observations of the sample mean are an estimate of the sampling distribution of the mean for samples of size 9 from this 'parent' population.

Theory says that mean, standard deviation, and shape are expected to be, 163, $\frac{6}{\sqrt{9}} = 2$, and normal, respectively (n doesn't have to be large because the parent is normal). The following procedure will make Minitab for Windows carry out the simulation, calculate 200 sample means in C10 and calculate the mean and standard deviation of the distribution of the sample mean, and draw a sketch of its shape:

Choose **Calc** > **Random data** > **normal**

Enter **200** in **Generate** box

Enter **C1-C9** in **Store in cols** box

Enter **163** in **mean** box

Enter **6** in **s.d.** box

Choose **OK**

If at this stage, you want to check that the simulation has been carried out correctly:

Choose **File** > **Display Data**

Enter **C1-C9** in the **Cols. to display** box.

Choose **OK**

You should see a table consisting of 200 rows and 9 columns, in which each cell in the table contains someone's height. Then:

Choose **Calc** > **Mathematical Expressions**

Enter **C10** in **Variable** box

Enter **(C1 + C2 + C3 + C4 + C5 + C6 + C7 + C8 + C9)/9** in **Expressions** box

Choose **Stat** > **Basic Statistics** > **Descriptive Statistics**

Enter **C10**

Choose **OK**

Choose **Graph** > **Character Graphs** > **Histogram**

Enter **C10**

Choose **OK**

Choose **File** > **Print Window**

Look at the Mean and 'Stdev' for C10. They should be close to 163 and 2, respectively. For example, I got 162.82 and 1.96, which seem reasonably close to the theoretical values. Also look at the shape of the histogram, which should be normal.

The reader is urged to try:

1. Varying the number of samples selected (200 in the example above) for the same parent;
2. Varying the sample size (9 in the example) for the same parent;
3. A nonnormal parent, for example a Poisson distribution with a mean of 0.5.

8.7 Summary

A population is the whole set of measurements or counts about which we want to draw a conclusion, and a sample is a subset of a population. Conclusions may be drawn from sample data if an appropriate sampling method and sample size are chosen. Random sampling is important because the theory of how to make inferences about populations from randomly sampled data is well developed. Systematic sampling may sometimes be used instead of random sampling, and stratified sampling is appropriate when the population is, in reality, made up of a number of subpopulations. The required sample size depends on the precision required in the estimate of the population parameter, and on the variability in the population.

The theory of inference depends on the distribution of sample statistics, such as the sample mean. The theory relates the characteristics of the sampling distribution to those of the parent population.

It is possible to simulate the sampling distribution of the sample mean using Minitab for Windows.

Worksheet 8:   Samples and Populations

Questions 1 to 5 are multiple choice. Choose one of the three options in each case:

1. The distribution of the means of samples of size 4, taken from a population with a standard deviation σ, has a standard deviation of: (a) $\frac{\sigma}{4}$, (b) σ, (c) $\frac{\sigma}{2}$.
2. A random sample is one in which:
   (a) Some members of the population are more likely to be picked than others.
   (b) Each member of the population has an equal chance of being picked.
   (c) Members of the population are picked because they are thought to be representative.
3. Conclusions drawn from sample data about populations are always subject to uncertainty because:
   (a) Data are not reliable,
   (b) Calculations are not accurate,
   (c) Only part of the population data are available.
4. The purpose of statistical inference is
   (a) To draw conclusions about populations from sample data,
   (b) To draw conclusions about populations and then collect sample data to support the conclusions,
   (c) To collect sample data and use them to formulate hypotheses about a population.
5. A random sample is preferred by statisticians because it:
   (a) Is representative of the population,
   (b) Ensures against bias,
   (c) Is chosen in a special statistical way.
6. What is
   (a) A population?
   (b) A sample?
   (c) A random sample?
   (d) A biased sample?
   (e) A census?
7. Why and how are random samples taken? Think of an example in your main subject area of interest. If nonrandom samples are taken, why are they? Do you think inferences from such samples are of value?

8. What is wrong with each of the following methods of sampling? In each case think of a better method.

   (a) In a laboratory experiment with mice, 5 mice were selected by the investigator by plunging a hand into a cage containing 20 mice and catching them one at a time.

   (b) In a survey to obtain adults' views on unemployment, people were stopped by the investigator as they came out of (i) A travel agent, (ii) A food supermarket, (iii) A job centre.

   (c) In a survey to decide whether two species of plant tended to grow close together in a large meadow, the investigator went into the meadow and randomly threw a quadrat over his left shoulder a number of times, each time noting the presence or absence of each species (a quadrat is, typically, a metal or plastic frame one metre square).

   (d) A survey of the price of bed and breakfast was undertaken in a town with 40 three-star, 50 two-star, and 10 one-star hotels. A sample of 10 hotels was obtained by numbering the hotels from 00 to 99 and then using random number tables.

   (e) To save time in carrying out an opinion poll in a constituency, a party worker selected a random sample from the constituency's electoral register and telephoned those selected.

   (f) To test the efficacy of an anti-influenza vaccine in his practice, a GP placed a notice in his surgery asking patients to volunteer to be vaccinated.

9. Answer this question using Minitab only.

   When a die is thrown once, the probability distribution for the number on the uppermost face is

Number	1	2	3	4	5	6
Probability	1/6	1/6	1/6	1/6	1/6	1/6

It can be shown that the mean and standard deviation of this distribution are 3.5 and 1.71. Now simulate the following experiment. Throw two dice and calculate the mean of the two numbers on the uppermost face. We will refer to this mean as the **'score'**. Repeat this until you have completed 108 throws of the dice. Obtain a histogram for the 'score', noting that the possible values for the score are 1, 1.5, 2, 2.5,..., 5.5, 6. Do you think the distribution is normal? If not, why not?

Also estimate the mean and standard deviation of the score and compare your answers with the theoretical values.

Repeat this simulation experiment for three dice.

Note: The distribution of a **single** throw of a die is given in the table above in Question 9. It can be described either as a **discrete** or as an **integer** distribution when we use Minitab. Although the integer distribution is restrictive since it allows only integer (whole number) values for the variable, and the probabilities must all be equal, this is exactly what we want here(!), so you are advised to use 'integer' instead of 'discrete' when you specify the the distribution of the parent population.

10. This question is typical of the first parts of a Statistics project which might have been undertaken by a student studying Statistics (together with another subject, for example Economics, for a joint honours degree) on the famous (!) Modular Course at Oxford Brookes University. What follows is intended to highlight the need to do a lot of thinking before you start to collect any data in a project.

    Suppose the aim of this project was to discover whether full-time undergraduate students at the university, who had taken paid jobs of at least 5 hours per week during term time, performed as well academically as similar students who had not taken such jobs. Suppose a student who undertook this project sent a questionnaire via e-mail to 1 in 10 of current undergraduates by taking the top 10% of names from the alphabetical list of full-time students held by the administration office of the university. Since there were 8000 full-time students in all, the 'project' student expected that about 800 questionnaires had been sent out. He noted that the date was February 14th, 1999. The project report was due to be completed by the end of April.

    That is all the information you will receive about this student and his project at this stage. However, you may find the following information, which relates to the general organisation of the Modular Course:

    The academic year consists of three terms and each module lasts for one term and is assessed during the term by coursework and/or by an end of term examination. You may assume that, if paid work has been undertaken during a particular term, it will at most affect the student's academic performance in that term only. Full-time students are required to pass 9 modules per year over a period of 3 years, making 27 modules in total. The average mark in the best 18 modules passed in the second and third years determines the degree classification.

    Please answer the following questions mainly concerning the work the student has done so far on his/her project:
    (a) What questions would you have included in the questionnaire? Make a list of all your questions.

(b) Why do you think the student chose a sample size of 800? How many would you have included? Give reasons for your answer.

(c) Do you agree with project student's method of selecting the sample? Give reasons for your answer. If you would have used a different sampling method, give details and a rationale.

(d) Assuming, for the sake of argument, the correct sampling method is used, what might be unsatifactory with the number of replies the student receives? What can be done to improve the situation?

(e) What information would you extract from the completed questionnaires? How would you present this information graphically? What summary statistics would you use in presenting the data from the questionnaire?

(f) If the average marks for the two groups of students, i.e., those who undertook paid term-time jobs and those who didn't, differed by:
  (a) 1%, what would you conclude?
  (b) 5%, what would you conclude?
  (c) 15%, what would you conclude?
  Assuming that you haven't given the same answer to all three questions, what made you change your mind?

(g) Is there a case for carrying out a pilot project or survey? What is implied by the word 'pilot'?

Notes:

1. It is a little unfair of me (!) to ask you some of the questions above, particularly part (f), because you have presumably read only to the end of Chapter 8. You will find the questions easier to answer when you have completed the whole book, or at least up to and including Chapter 11.

2. Some of the questions concerning this project do not have a correct answer, possibly because you do not have all the information you need. Just do your best. Good Luck!

Chapter 9

Confidence Interval Estimation

9.1 Introduction

As stated in Section 8.1, the remainder of this book is concerned with **Statisical Inference**, which can be divided into two topics, namely: **Confidence Interval Estimation** and **Hypothesis Testing**. These topics are the titles of Chapters 9 and 10, respectively, and (as we shall see in Chapter 10) they are, in fact, connected.

Consider the random sample of 27 female student heights recorded in Table 1.1. These heights are represented graphically in part of Fig. 3.10, reproduced here as Fig. 9.1.

By calculation we find that the sample mean height is $\bar{x} = 163.4$ cm, but what is our estimate of the population mean height μ (i.e., of all first-year female students in U.K. universities)? The obvious answer is 163.4 cm, if we require a single-value estimate (often referred to as a **point** estimate). However, since our estimate is based on a sample of the population of heights, we might want to be more guarded and give some idea of the precision of the estimate by adding and subtracting an **'error' term** which might, for example, lead to a statement that our estimate is 163.4 $\pm$ 1, meaning that the population mean height lies between 162.4 and 164.4 cm. Such an estimate is referred to as an **interval** estimate. Statistical theory indicates that the size of the error term and hence the **width** of the interval, depend on

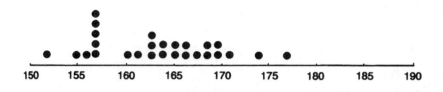

Figure 9.1   Dotplot of the Heights of 27 Female Students

three factors:

1. The **sample size, _n_**—the larger the sample size, the smaller the error term and the smaller the width of the interval (other things being equal).
2. The variability of height (as measured by the standard deviation) — the more variable the height, the larger the standard deviation, the greater the error term, and the greater the width of the interval.
3. The level of confidence we wish to have that the population mean height does in fact lie within the specified interval — the greater the confidence, the greater the error term and the greater the width of the interval.

This third factor, confidence, is such an important concept that we will devote the next section to it.

Note

The three statements above are quoted without proof. I hope that you can at least accept them as being 'intuitively reasonable'.

9.2   95% Confidence Intervals

For the student height example, the population mean height, μ, has a fixed numerical value at any given time. This value is unknown to us, but by taking a random sample of 27 heights, we want to specify an interval within which we are reasonably confident that this fixed unknown value lies.

Suppose we decide that nothing less than 100% confidence will suffice, implying absolute certainty. Unfortunately, theory indicates that the 100% confidence interval is so wide that it is useless for all practical purposes.

Instead, statisticians conventionally choose a **95% confidence level** and calculate **a 95% confidence interval** for the population mean, μ, using formulae we shall introduce in the next sections. For the moment, it is important for you to understand what a confidence level of 95% means. It means that on 95% of occasions when such intervals are calculated the population mean will actually fall inside the interval we have calculated from the sample data. On the other 5% of occasions, it will fall outside the interval. However, in a particular case, we do not know whether the mean has been successsfully 'captured' within the calculated interval.

9.3 Calculating a 95% Confidence Interval for the Mean, μ, of a Population: Large Sample Size

Having discussed some concepts, we now discuss methods of actually calculating a 95% confidence interval for the population mean.

In Section 8.5, we stated that the sample mean, $\bar{x}$, has a distribution with mean μ and standard deviation $\frac{\sigma}{\sqrt{n}}$, and is approximately normal if n is large. Recall that μ and σ are the mean and standard deviation of the distribution of the 'parent' population. It follows from the theory of the normal distribution (refer to Section 7.2 if necessary) that, for example, '95% of the distribution of the sample mean lies within 1.96 standard deviation of the mean of the distribution of the sample mean'. Not only is this statement 'a bit of a mouthful', it is probably 'indigestible' if you are seeing it for the first time! Fig. 9.2 is intended to make the statement more digestible.

So, we can state that 95% of sample means lie in the range:

$$\bar{x} - 1.96\frac{\sigma}{\sqrt{n}} \quad \text{to} \quad \bar{x} + 1.96\frac{\sigma}{\sqrt{n}}$$

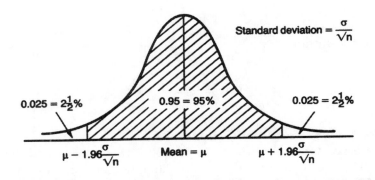

Figure 9.2 Distribution of the Sample Mean $\bar{x}$, if n is Large

We can write this as a probability statement:

$$P\left(\mu - 1.96\frac{\sigma}{\sqrt{n}} < \bar{x} < \mu + 1.96\frac{\sigma}{\sqrt{n}}\right) = 0.95$$

This statement can be rearranged to state:

$$P\left(\bar{x} - 1.96\frac{\sigma}{\sqrt{n}} < \mu < \bar{x} + 1.96\frac{\sigma}{\sqrt{n}}\right) = 0.95$$

If we now use the fact that n is large, we can replace σ by the sample standard deviation, s, which we calculate from sample data, along with $\bar{x}$, we obtain the following statement, which is approximately true:

$$P\left(\bar{x} - 1.96\frac{s}{\sqrt{n}} < \mu < \bar{x} + 1.96\frac{s}{\sqrt{n}}\right) = 0.95$$

This result is important because we can calculate

$$\bar{x} - 1.96\frac{s}{\sqrt{n}} \quad \text{and} \quad \bar{x} + 1.96\frac{s}{\sqrt{n}}$$

from our sample data. These two values are called the **95% confidence limits** for μ, and are usually written as:

$$\bar{x} \pm 1.96\frac{s}{\sqrt{n}} \tag{9.1}$$

The interval from

$$\bar{x} - 1.96\frac{s}{\sqrt{n}} \quad \text{to} \quad \bar{x} + 1.96\frac{s}{\sqrt{n}}$$

is called the **95% confidence interval** for μ.

The **error term** (referred to in Section 9.1) is $1.96\frac{s}{\sqrt{n}}$ and the **width** of the 95% confidence interval is the difference between the two limits. Remember that we can only use the formula in this section when n is large (greater than 30, say).

Note

Some texts discuss the case where σ, the standard deviation of the population, is known, i.e., as an exact value. This author's view is that this is an unrealistic case, since, if we know σ we must have all the measurements in the population. Hence we can also calculate the population mean, μ, exactly, so there is no need to estimate it.

Example

Suppose that, from a random sample of 27 heights, we calculate that the sample mean is 163.4 cm, and the sample standard deviation is 6.1 cm (these summary statistics apply to the data referred to in Section 9.1). Then a 95% confidence interval for the population mean height, μ, is

$$163.4 - \frac{1.96 \times 6.1}{\sqrt{27}} \quad \text{to} \quad 163.4 + \frac{1.96 \times 6.1}{\sqrt{27}}$$

$$163.4 - 2.3 \quad \text{to} \quad 163.4 + 2.3$$

$$161.1 \quad \text{to} \quad 165.7$$

So we are 95% confident that μ lies between 161.1 and 165.7 cm. Also, the error term is 2.3 cm, and the width of the interval is $165.7 - 161.1 = 4.6$ cm, twice the error term (see Fig. 9.3).

The formula for the error term is $1.96 \frac{s}{\sqrt{n}}$, and this can be used to support the intuitive arguments of Section 9.1 concerning the three factors which affect the error term:

1. Clearly, as n increases, $1.96 \frac{s}{\sqrt{n}}$ decreases.
2. The larger the variability, the larger s will be, so $1.96 \frac{s}{\sqrt{n}}$ will be larger.
3. We use the factor 1.96 in the error term, if our confidence level is 95%. For greater confidence we would need to move further away from the mean (see Fig. 9.2) so the factor 1.96 would increase and so would the error term. So, the greater the confidence level, the greater the error term, and the wider the interval.

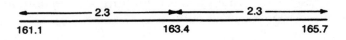

Figure 9.3   A 95% Confidence Interval for μ

9.4 Calculating a 95% Confidence Interval for the Mean, μ, of a Population: Small Sample Size

The formula of the previous section referred to cases where the sample size was large. This can be taken to mean 'greater than 30' as a rough guide. If we can reasonably assume that the variable of interest, x, is approximately normally distributed, a 95% confidence interval for the population mean is given by:

$$\bar{x} - \frac{ts}{\sqrt{n}} \quad \text{to} \quad \bar{x} + \frac{ts}{\sqrt{n}} \quad \text{for} \quad n > 1$$

which is usually written

$$\bar{x} \pm \frac{ts}{\sqrt{n}} \tag{9.2}$$

The value of t is obtained from Table C.5 of Appendix C and depends on two factors:

1. The confidence level, which determines the value of α to be entered in Table C.5. For example, for 95% confidence,

$$\alpha = \frac{1 - 0.95}{2} = 0.025.$$

2. The sample size, n, which determines the number of degrees of freedom, ν (Greek letter nu) to be entered in Table C.5. The general idea of what is meant by the term 'degrees of freedom' is discussed in Section 9.7. In using Formula (9.2), $\nu = (n - 1)$. Note also that the error term in Formula (9.2) is $\frac{ts}{\sqrt{n}}$, and the width of the interval is $\frac{2ts}{\sqrt{n}}$.

We stated earlier in this section that the variable of interest is approximately normal. However, this assumption is less critical the larger the value of n. The difficulty with this assumption arises when n is small, say below 15. How can you tell whether a small sample has been taken from a normal distribution if all you have are a few observed values? You can, I suppose, draw a dotplot and look for two characteristics of any normal distribution, namely, **symmetry** and **bunching in the middle**. Clearly, the following is approximately normal:

But what about the next dotplot? It exhibits symmetry, but not bunching.

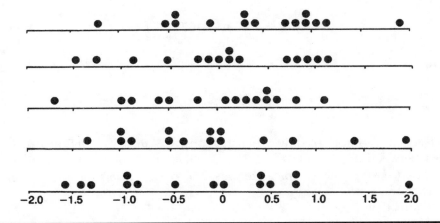

And what about the third dotplot? It exhibits bunching, but not symmetry.

Here is a suggestion! Use the Minitab simulation method of Section 7.5 to take a sample of 15 observations from any normal distribution, for example, $N(0,1)$, and draw a dotplot. Repeat this for several other random samples of the same size from the same distribution to get a 'feel' for what small random samples taken from a normal distribution look like. Fig. 9.4 shows five dotplots from the $N(0,1)$ distribution.

Example

Suppose we take a sample of only nine female student heights and obtain the following observations:

163, 157, 160, 168, 155, 168, 164, 157, 169.

From these data, $\bar{x} = 162.3$, $s = 5.3$. A dotplot for the data is given in Fig. 9.5.

Figure 9.4 Dotplots of Five Samples of Size 15 Taken from $N(0,1)$

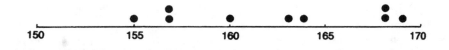

Figure 9.5   Dotplot of the Heights of Nine Female Students

Given Fig. 9.5, I would be quite happy to assume approximate normality, not only because the dotplot looks reasonable but also because the heights here are of a sample from a population of nominally identical individuals (refer to the first few lines of Section 7.2). Nevertheless, this is a subjective decision. If you are not happy to assume normality yet, there are two things you can do:

1. Carry out a formal test of normality (to be discussed in Section 16.5).
2. Take a larger sample, both to give a better plot and also to reduce the importance of the assumption of normality.

We now need to find a value for t, to be used in Formula (9.2). For 95% confidence, $\alpha = 0.025$, $\nu = 9 - 1 = 8$ when $n = 9$. Hence, from Table C.5, $t = 2.306$. So a 95% confidence interval for μ is

$$162.1 \pm \frac{2.306 \times 5.3}{\sqrt{9}},$$

or

$$162.1 - 4.1 \quad \text{to} \quad 162.1 + 4.1,$$

or

$$158.0 \quad \text{to} \quad 166.2$$

Based on a sample of only 9 observations, we are 95% confident that μ lies between 158.0 and 166.2. The width of the interval has increased from 4.6 (when the sample size was 27) to 8.2 (when the sample size was only 9).

Note that if we use Table C.5 for large values of n, then for 95% confidence and hence $\alpha = 0.025$, the t value rapidly approaches 2, and for very large samples is 1.96, and Formulae (9.2) and (9.1) are the same!

It follows that we may use Formula (9.2) for all n greater than 1 (and stop wondering what 'n must be large' means). However, we must still be able to justify approximate normality if n is small.

Minitab for Windows can be used to obtain a 95% confidence interval for μ (see Table 9.1) which follows:

Table 9.1   A 95% Confidence Interval for μ; the Minitab Input and Output

---

Input

Enter the 9 height values into column C1, say.
Choose **Stat** > **Basic Stats** > **1-Sample t**
Enter **C1** in **Variables** box
Choose **Confidence interval**
Enter **95** in **Level** box.
Choose **OK**

Output

C1
  163   157   160   168   155   168   164   157   169
MTB > Tinterval 95.0 C1.

	N	MEAN	STDEV	SE MEAN	95.0 PERCENT C.I.
C1	9	162.33	5.34	1.78	(158.23, 166.44)

9.5   The *t* Distribution

This continuous probability distribution was first studied by W.S. Gosset, who published his results under the pseudonym of 'Student', which is why the distribution is often referred to as **Student's *t* distribution**. It arises when we consider taking a large number of random samples of the same size, n, from a normal distribution with known mean, μ. Then the probability distribution of the statistic:

$$ t = \frac{\bar{x} - \mu}{\frac{s}{\sqrt{n}}} $$

may be plotted. It will be symmetrical and unimodal. For different values of n, different distributions will be obtained; for large n the t distribution approaches the standardized normal distribution, $N(0,1)$,

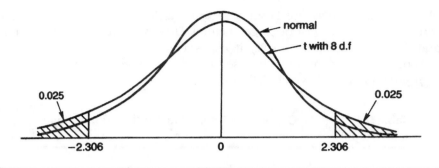

Figure 9.6 Comparison of the Shapes of a Normal Distribution and a *t* Distribution with $\nu = 8$ Degrees of Freedom

while for small n the t distribution is flatter and has higher tails than $N(0,1)$ (see Fig. 9.6).

9.6 The Choice of Sample Size When Estimating the Mean of a Population

In Section 8.4, we discussed the choice of sample size, n, but deferred deciding how to calculate how large it should be. Instead we concentrated on the factors affecting the choice of n for the case of estimating μ, the mean of a population. These factors were

1. The precision with which the population mean is to be estimated, and we can now state this precision in terms of the 'error' term in the Formula (9.2) for the confidence interval for μ.
2. The variability of the measurements, and we noted a chicken-and-egg situation of needing to know the variability before we had any sample data. In order to overcome this difficulty either carry out a small pilot experiment or use estimates of standard deviation from your own or another researcher's relevant work.

Example

Suppose that in estimating the mean, we specify an error term of 1 and a confidence level of 95%. Then we know that:

$$\frac{ts}{\sqrt{n}} = 1$$

where t is found from Table C.5 for $\alpha = 0.025$, but n and hence $(n - 1)$ are unknown. Suppose further that we also have a rough estimate, from a small **pilot experiment** that $s = 10$. Now we can state that:

$$\frac{t \times 10}{\sqrt{n}} = 1$$

How can we solve this equation, since t depends on the value of n? The trick is to assume that n is large, and note that, for $\alpha = 0.025$, t is roughly 2 for large values of n. Now we can solve:

$$\frac{2 \times 10}{\sqrt{n}} = 1$$

by squaring both sides of the equation, giving $n = 400$. (We were correct in assuming n would be large.)

9.7 Degrees of Freedom

There are two approaches which you, the reader, can take to the concept of **degrees of freedom**. The 'surface' approach is to know where to find the formula for calculating degrees of freedom for each application covered (and there are several in the remaining chapters). The 'in-depth' or more mature approach is to try to understand the general principle behind all the formulae for degrees of freedom (d.f. for short):

> The number of d.f. may be defined as **'the number of independent observations employed in calculating a statistic which will be used in estimation or hypothesis testing, minus the number of restrictions placed on the sample data'.**

Example

Why do we use $(n - 1)$ d.f. when we look up t in Table C.5 as part of the calculation of a 95% confidence interval for the population mean? The answer is that in the Formula (9.2), we caluclate the standard deviation, s, using Formula (4.3), which involves summing the squares of the deviations of the n sample observations from the sample mean. It would appear that we have n independent observations. Yes, we have; BUT we have only $n - 1$ independent deviations from the mean since we know that 'the sum of the n deviations from the mean is always zero'. Remember $\Sigma(x - \bar{x}) = 0$? It is a result that was mentioned several times, for example, in Section 2.1, in Worksheet 2, and also in Section 4.8. So, once we know $n - 1$ of the deviations from the mean, the other deviation must be such that the sum of all n deviations is zero.

9.8 95% Confidence Interval for a Binomial Probability

The discussion so far in this chapter has been concerned with confidence intervals for the mean of a population. If our sample data are from a binomial experiment for which we do not know the value of the parameter, p, the probability of success in each trial (in other words, the proportion of successes in a large number of Bernoulli trials), then we can use our sample data to calculate a 95% confidence interval for p. For example, if we observe x successes in the n trials of a binomial experiment, a 95% confidence interval for p is

$$\frac{x}{n} \pm 1.96\sqrt{\frac{\frac{x}{n}(1 - \frac{x}{n})}{n}} \tag{9.3}$$

provided $x > 5$ and $(n - x) > 5$. These two conditions are the equivalent of $np > 5$ and $np(1 - p) > 5$ for the 'normal approximation to the binomial' (Section 7.7), where the unknown p is replaced by its point estimator, $\frac{x}{n}$.

Note also that in using this formula, the four conditions for the binomial must apply (see Section 6.3).

Example

Of a random sample of 200 voters taking part in an opinion poll a few days before an election, 110 said they would vote for party A, the other 90 said they would vote for other parties. What proportion of the total electorate will vote for party A?

If we regard 'voting for A' as a 'success', then $x = 110$, $n = 200$. The conditions $x > 5$ and $(n - x) > 5$ are satisfied, so a 95% confidence interval for p is

$$\frac{110}{200} \pm 1.96 \sqrt{\frac{\frac{110}{200}\left(1 - \frac{110}{200}\right)}{200}}$$

$$0.55 \pm 0.07$$

$$0.48 \quad \text{to} \quad 0.62$$

We can be 95% confident that the proportion who will vote for party A is between 0.48 (48%) and 0.62 (62%). Of course, we have to make a number of assumptions in this kind of survey, not least of which is that the voters told the truth about their voting intentions, and did not change their minds between the poll and the election.

9.9 The Choice of Sample Size When Estimating a Binomial Probability

In the example of the previous section the width of the confidence interval is quite large. If we wished to reduce the width by reducing the error term, one way of doing this is by increasing the sample size.

Example

If we wished to estimate the proportion to 'within an error term of 0.02' for a confidence level of 95%, the new sample size, n, could be found by solving the equation:

$$1.96 \sqrt{\frac{0.55(1 - 0.55)}{n}} = 0.02$$

Squaring both sides gives

$$1.96^2 \times \frac{0.55 \times 0.45}{n} = 0.02^2$$

So,

$$n = 1.96^2 \times \frac{0.55 \times 0.45}{0.02^2} = 2377$$

We need a sample of nearly 2500. Notice how we have again used the result of a **pilot survey** (of 200 voters), as in Section 9.6.

9.10 95% Confidence Interval for the Mean of a Population of Differences: 'Paired' Samples Data, and Including Minitab

In experimental work we are often concerned with not just one population, but with a comparison between two populations. For example, suppose that two methods of teaching children to read are to be compared. Some children are to be taught by a standard method (S), while the rest are to be taught by a new method (N). In order to reduce the effect of factors other than the teaching method, children are matched in pairs so that children in each pair are as similar as possible with respect to factors such as age, sex, social background, and initial reading ability. One child from each pair is then **randomly assigned** to teaching method S, and the other in the pair to method N.

Suppose that after one year the children are tested for reading ability, and that the data in Table 9.2 are the test scores for 10 pairs of children.

In this example we can think of **two** populations of measurements, namely, the S method scores and the N method scores. However, our main interest is in the **difference** between the scores obtained by the two methods. For example, for the first pair in Table 9.2 this difference is 7. The 10 values in the bottom row of the table are, in fact, a sample from **one** population, namely, the **population of differences,** d, say, in the scores obtained by the two methods.

The sample data in Table 9.2 are an example of what is often referred to as **paired** samples data. The differences, d, in the bottom row of Table 9.2, have been calculated using the formula: $d = N$ score - S score. This implies that high positive values of d will tend to support the idea that method N is better than method S, and vice versa.

Table 9.2 Reading Test Scores of 10 Matched Pairs of Children

Pair Number	1	2	3	4	5	6	7	8	9	10
S method score	56	59	61	48	39	56	75	45	81	60
N method score	63	57	67	52	61	71	70	46	93	75
$d = N$ score — S score	7	−2	6	4	22	15	−5	1	12	15

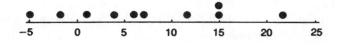

Figure 9.7 Dotplot for the Differences d in Table 9.2

A 95% confidence interval for μ_d, the **mean of the population of differences**, is given by Formula (9.4):

$$\bar{d} \pm \frac{ts_d}{\sqrt{n}} \qquad (9.4)$$

where $\bar{d}$ and s_d are the mean and standard deviation, respectively, of the sample of differences, so:

$$\bar{d} = \frac{\Sigma d}{n} \quad \text{and} \quad s_d = \sqrt{\frac{\Sigma d^2 - \frac{(\Sigma d)^2}{n}}{n-1}} \qquad (9.5)$$

In Formulae (9.4) and (9.5), n stands for the number of differences (= number of pairs). Hence, the value of t is obtained from Table C.5 for $\alpha = 0.025$ and $\nu = (n - 1)$. In order to calculate a 95% confidence interval for μ_d we must be able to assume that the **differences are approximately normally distributed**. This assumption is less critical the larger the value of n.

A dotplot for the data in Table 9.2 is shown in Fig. 9.7. The differences do not seem to be markedly nonnormal.

From the 10 values of d, we calculate $\bar{d} = 7.5$, $s_d = 8.48$, using Formula (9.5). We also know that $n = 10$ and hence $t = 2.262$ from Table C.5. So a 95% confidence interval for μ_d is

$$7.5 \pm \frac{2.262 \times 8.48}{\sqrt{10}}$$

using formula (9.4), i.e., 1.4 to 13.6.

We are 95% confident that μ_d, the mean of the population of the difference in the scores from the two methods, lies between 1.4 and 13.6. A tentative conclusion at this stage is that method N gives higher scores on average than method S. A more formal conclusion will be given in the next chapter.

As in Section 9.6, it would now be possible to decide what sample size to choose in another experiment designed to provide a more precise estimate of the mean difference in scores between the two teaching methods.

In order to use Minitab for Windows to obtain a 95% confidence interval for the mean, simply follow the method of Table 9.1, given earlier in this chapter, by entering the 10 differences in C1. Alternatively, enter method S scores into C1 and the method N scores into C2, then follow the steps of Table 9.3 as follows:

Table 9.3   Confidence Interval for μ_d, Paired Samples Data

Choose **Calc** > **Mathematical Expressions**
Enter **C3** in the **Variable** box
Enter **C2 - C1** in the **Expression** box
Choose **Stat** > **Basic Stats** > **1- Sample** *t*
Enter **C3** in **Variables** box
Choose **Confidence interval**
Enter **95** in the **Level** box
Choose **OK**

You should get the same answers (1.4 to 13.6) as above for the 95% C.I.

9.11   95% Confidence Interval for the Difference in the Means of Two Populations, 'Unpaired' Samples Data, and Including Minitab

The example of the previous section was not, in essence, a comparison of two populations since the data were in pairs. In many other instances in which two populations of measurements are concerned, the data are **unpaired**.

For example, the A-level counts of a random sample of 40 students studying for a BA or BSc degree were summarised as shown in Fig. 9.8 from the data given in Columns 6 and 7 of Table 1.1.

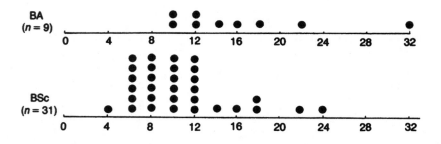

Figure 9.8   Dotplot for the A-level Counts of BA and BSc Students

Suppose we wish to calculate a 95% confidence interval for $(\mu_1 - \mu_2)$, the difference in the means of the two populations. Here μ_1 is the population mean A-level count for BA students, and μ_2 is the population mean A-level count for BSc students. Note that the data used to draw Fig. 9.8 are unpaired in the sense that no A-level count in the first sample (BA students) is associated with any particular A-level count in the second sample (BSc students).

The formula for a 95% confidence interval for $(\mu_1 - \mu_2)$ is

$$(\bar{x}_1 - \bar{x}_2) \pm ts \sqrt{\frac{1}{n_1} + \frac{1}{n_2}} \qquad (9.6)$$

where $\bar{x}_1$ is the sample mean A-level count for the first sample of size n_1 and $\bar{x}_2$ is the sample mean A-level count for the second sample of size n_2. Also, t is found from Table C5 for $\alpha = 0.025$, $\nu = (n_1 + n_2 - 2)$ degrees of freedom, and s^2 is given by Formula (9.7):

$$s^2 = \frac{(n_1 - 1)s_1^2 + (n_2 - 1)s_2^2}{n_1 + n_2 - 2} \qquad (9.7)$$

In fact, s^2 is a weighted average of the two sample variances s_1^2 and s_2^2 and is called a **pooled** estimate of the common variance of the two populations (see assumption 2 below).

In using Formula (9.7), we must make the two assumptions which follow:

1. The measurements in each population must be approximately normally distributed, this assumption being less critical the larger the values of n_1 and n_2.
2. The population variances, σ_1^2 and σ_2^2 (and hence the standard deviations) must be equal.

For the numerical example of the A-level count data, the first sample size, $n_1 = 9$, is very small, but the dotplot (Fig. 9.8) is not obviously non-normal. For the second sample, $n_2 = 31$, so the assumption of normality is less important. We observe some bunching, but there is also an indication of positive skewness. However, the coefficient of skewness (Section 4.13) is only 0.45. As this is less than 1, we can reasonably conclude that the skewness is not marked.

The second assumption requires $\sigma_1^2 = \sigma_2^2$, i.e., $\sigma_1 = \sigma_2$. One way of deciding whether this is a reasonable assumption in this case is to look at s_1 and s_2, the sample standard deviations. For our data, $s_1 = 7.10$ and $s_2 = 4.78$. This is not such good agreement, but is it good enough?

The answer is 'yes', because an 'F test' indicates that the hypothesis that $\sigma_1^2 = \sigma_2^2$, should not be rejected. (Hypothesis tests are introduced in Chapter 10 and the F test is introduced in Section 10.15)

In order to carry out the calculations for a 95% confidence interval for $(\mu_1 - \mu_2)$, we note the following summary statistics:

$$\bar{x}_1 = 16.22 \quad \bar{x}_2 = 10.71$$
$$s_1 = 7.10 \quad s_2 = 4.78$$
$$n_1 = 9 \quad n_2 = 31$$

Then we obtain s using Formula (9.7):

$$s^2 = \frac{(9-1) \times 7.10^2 + (31-1) \times 4.78^2}{9 + 31 - 2}$$
$$s^2 = 28.5$$
$$s = \sqrt{28.65} = 5.35$$

So for a 95% confidence interval for $(\mu_1 - \mu_2)$ using Formula (9.6)

$$16.22 - 10.71 \pm 2.02 \times 5.35 \sqrt{\frac{1}{9} + \frac{1}{31}}$$
$$5.51 \pm 4.09$$
$$1.4 \quad \text{to} \quad 9.6$$

where t is taken from Table C.5 for $\alpha = 0.025$, $\nu = 9 + 31 - 2 = 38$.

We are 95% confident that the difference between the population mean A-level counts of BA and BSc students is between 1.4 and 9.6. We note that the confidence interval does not contain the value zero, since confidence limts, 1.4 and 9.6, have the same sign. Both limits being positive indicate that the mean A-level count for BA students is '**significantly higher**' than for BSc students. We will return to this important idea in the next chapter.

As in previous examples in this chapter, more precise estimates of the difference in the means could have been obtained by taking larger sample sizes. Minitab for Windows can be used to obtain a 95% confidence for $(\mu_1 - \mu_2)$. For the example above, there are two ways of setting up the data, which we will refer to as 'Stacked' and 'Unstacked'. We note that the relevant data for A-level count and Type of degree are stored in columns 6 and

7 of a Minitab file called ES4DAT.MTW, which you will have access to if you have already typed it into your computer (see Section 3.2).

Table 9.4   Two Methods for Finding a Confidence Interval for
$(\mu_1 - \mu_2)$**, Unpaired Samples Data, Stacked and Unstacked**

Stacked

Choose **Stat** > **Basic Stats** > **2-Sample** *t*
Choose **Samples in one column**
Enter **C6** in **Samples** box
Enter **C7** in **Subscripts** box
Choose **'not equal to'** in the **Alternative** box
Enter **95** in **Confidence level** box
Choose **Assume equal variances** by clicking in the little box
Click on **OK**
    (For output see below)

Unstacked

Starting with a new, i.e., blank, spreadsheet,
Enter the 9 A-level counts for the 9 BA students in C1
Enter the 31 A-level counts for the 31 BSc students in C2
Choose **Stat** > **Basic Stats** > **2-Sample** *t*
Choose **Samples in different columns**
Enter **C1** in **'First'** box
Enter **C2** in **'Second'** box
Choose **'not equal to'** in the **Alternative** box
Enter **95** in the **Confidence level** box
Choose **Assume equal variances** by clicking in the little box
Click on **OK**
    (For output see below)

Output (this is the same for the two methods—stacked or unstacked)

TWO SAMPLE *T* FOR BA VS BSc

	N	MEAN	STDEV
BA	9	16.22	7.10
BSc	31	10.71	4.77

95 PCT CI FOR MU BA - MU BSc: (1.4 , 9.6) POOLED STDEV = 5.35

Note that the confidence interval agrees with that found earlier in this section.

9.12 Summary

A confidence interval for an unknown parameter of a population, such as the mean, is a range within which we have a particular level of confidence, such as 95%, that the parameter lies within it. If we have randomly sampled data we can calculate confidence intervals for various parameters using appropriate formulae from Appendix A, but it is important to check whether the required assumptions are valid in each case.

We can also decide sample sizes if we can specify the precision with which we wish to estimate the parameter, and if we have some measure of variability based on the results of a pilot experiment or survey.

Worksheet 9: Confidence Interval Estimation

1. Why are confidence intervals calculated?
2. A 90% confidence interval for the mean of a population is such that:
   (a) 10% of the values in the population lie outside it.
   (b) There is a 90% chance that it contains all the values in the population.
   (c) There is a 90% chance that it contains the mean of the population. Which of (a), (b), or (c) is the correct statement?
3. The larger the sample size, the wider the 95% confidence interval. True or false?
4. The more variation in the measurements, the wider the 95% confidence interval. True or false?
5. The Formulae (9.1) and (9.3) apply only to 95% confidence intervals. What formulae would you use if the confidence level was set at (a) 99% (b) 90%?
6. The higher the confidence level, the wider the 95% confidence interval. True or false?
7. What does the following statement mean: "I am 95% confident that the mean of the population lies between 10 and 12"?
8. Of a random sample of 100 customers who had not settled their accounts with an Electricity Board within one month of receiving them, the mean amount owed was £30 and the standard deviation was £10. What is your estimate of the mean of all unsettled accounts? Suppose that the Electricity Board wanted an estimate of the mean of all unsettled accounts to be within £1 of the true figure for 95% confidence. How many customers who had not settled their accounts would need to be sampled?

9. Refer to Question 17 of Worksheet 5, concerning the game of Patience. The data were the results of 500 games of Patience in 10 blocks of 50 games per block.
   (a) Using the results of the first 50 games only, calculate a 95% confidence interval for p, the probability of winning a game, making two assumptions which should be stated.
   (b) Repeat (a) for 100, 200, 300, and 500 games. Comment on the widths of the five confidence intervals.
   (c) How many games would be needed to give a 95% confidence for p to within ± 0.03?
   (d) Discuss the validity of the two assumptions made in (a) (which were also made in [b] and [c]).

10. The systolic blood pressure of 90 normal British males has a mean of 128.9 mm of mercury and a standard deviation of 17 mm of mercury. Assuming these are a random sample of blood pressures, calculate a 95% confidence interval for the population mean blood pressure.
    (a) How wide is the interval?
    (b) How wide would the interval be if the confidence level was raised to 99%?
    (c) How wide would the 95% confidence interval be if the sample size was increased to 360?
    Are your answers to (a), (b), and (c) consistent with your answers to Questions 3 and 6 above?

11. In order to estimate the percentage of pebbles made of flint in a given locality to within 1% for 95% confidence, a pilot survey was carried out. Of a random sample of 30 pebbles, 12 were made of flint. How many pebbles need to be sampled in the main survey?

12. The number of drinks sold from a vending machine in a motorway service station was recorded on 60 consecutive days. The results were as follows:

30	40	60	70	120	130	140	150	160	170
180	190	200	200	210	210	220	230	240	250
260	260	270	280	280	290	290	300	300	310
320	320	330	330	340	350	350	360	360	360
360	370	370	380	380	390	390	400	410	420
430	440	460	470	480	490	510	550	590	610

Ignoring any differences between different days of the week and any time-trend or seasonal effects, estimate the mean number of drinks sold per day in the long term.

13. Ten women recorded their weights in kilograms before and after dieting. Assuming that the women were randomly selected, estimate the population mean reduction in weight. What additional assumption is required, and is it reasonable here? The weights were

Before	89.1	68.3	77.2	91.6	85.6	83.2	73.4	84.3	96.4	87.6
After	84.3	66.2	76.8	79.3	85.5	80.2	76.2	80.3	90.5	80.3

14. The percentage of a certain element in an alloy was determined for 16 specimens using two methods, A and B. Eight of the specimens were randomly allocated to each method. The percentages were

Method A	13.3	13.4	13.3	13.5	13.6	13.4	13.3	13.4
Method B	13.9	14.0	13.9	13.9	13.9	13.9	13.8	13.7

Calculate a 95% confidence interval for the difference in the mean percentages of the element in the alloy for the two methods, stating any assumptions made.

15. The annual rainfall in centimetres in two English towns over a period of 11 years was as follows:

Year	Town A	Town B
1970	100	120
1971	89	115
1972	84	96
1973	120	115
1974	130	140
1975	105	120
1976	60	75
1977	70	90
1978	90	90
1979	108	105
1980	130	135

Estimate the mean difference in the annual rainfall for the two towns.

16. The actual weights of honey in 12 jars marked 452 g were recorded. Six of the jars were randomly selected from a large batch of brand A honey, and six were randomly selected from a large batch of brand B honey. The weights were

| Brand A | 442 | 445 | 440 | 448 | 443 | 450 |
| Brand B | 452 | 450 | 456 | 456 | 460 | 449 |

Estimate the mean difference in the weights of honey in jars marked 452 g for the two brands. Also estimate separately:
(a) The mean weight of brand A honey, and
(b) The mean weight of brand B honey.
Decide whether it is reasonable to suppose that the mean weight of honey from the brand A batch is 452 g, and similarly for brand B honey.

17. This question is designed to help you to understand more about confidence intervals using simulation on Minitab.

Confidence intervals are relatively simple to calculate, but what do they mean when we have calculated them? In the case of a 95% confidence interval for a population mean, μ, the answer is given in Section 9.2. Similar statements can be made for confidence intervals other than 95% and for parameters such as μ_d and $(\mu_1 - \mu_2)$.

To illustrate the concept involved we can, for example, use simulation to take a number of samples of size n from a normally distributed population with a known mean and standard deviation. For each sample we can then calculate a 95% confidence interval. If the formula used to calculate the 95% confidence intervals is correct, then we would expect that 95% of such intervals will 'capture' the population mean. By 'capture' we mean that the known value of the population mean lies inside the confidence interval.

Table 9.5 is an example of the steps required to make Minitab for Windows simulate 100 confidence intervals for μ, the population mean based on samples of size 9 taken from a normal distribution with a mean of 70 and standard deviation 3.

You should count how many of the 100 intervals contain the value 70, and compare it with the theory which indicates that, on average, 95 of the 100 are expected to do so.

Table 9.5   Simulation of 100 Confidence Intervals Based on 100 Samples of Size 9 Taken from a N $(70,3^2)$ Distribution, Using Minitab for Windows

Choose **Calc** > **Random data** > **Normal**
Enter **9** in **Generate** Box
Enter **C1-C50** in **Store in columns** box
Enter **70** in **Mean** box
Enter **3** in **Standard deviation** box
Click on **OK**
Choose **Calc** > **Random data** > **Normal**
Enter **9** in **Generate** box
Enter **C51-C100** in **Store in columns** box
Enter **70** in **Mean** box
Enter **3** in **Standard deviation** box
Click on **OK**
Choose **Stats** > **Basic Stats** > **1-Sample** *t*
Enter **95.0** in **Confidence level** box
Click on **OK**
Choose **File** > **Print Window**

Chapter 10

Hypothesis Testing

What tribunal can possibly decide truth in the clash of contradictory assertions and conjectures?

10.1 Introduction

Statistical inference is concerned with how we draw conclusions from sample data about the larger population from which the sample has been selected. In the Chapter 9, we discussed one branch of inference, namely estimation, particularly **confidence interval estimation**. Another important branch of inference is **hypothesis testing** (Fig. 10.1), which is the subject of much of the remainder of this book.

In this chapter we will consider again the five applications we looked at in Chapter 9 (see Sections 9.3, 9.4, 9.8, 9.10, and 9.11), but this time in terms of testing hypotheses about the various parameters. We end the chapter by discussing the connection between the two branches of inference (see the dashed line in Fig. 10.1).

The procedure for performing any hypothesis test can be set out in terms of seven steps:

1. Decide on a null hypothesis, H_0.
2. Decide on an alternative hypothesis, H_1.
3. Decide on a significance level.
4. Calculate the appropriate test statistic, using the sample data.
5. Find from tables the appropriate tabulated test statistic.

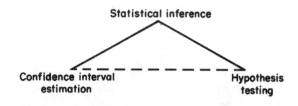

Figure 10.1 Types of Statistical Inference

6. Compare the calculated and tabulated test statistics, and decide whether to reject the null hypothesis, H_0.
7. State a conclusion, after checking to see whether the assumptions required for the test in question are valid.

Notes

The steps above apply mainly to hypothesis tests performed 'by hand', for example, with a calculator and/or in an examination. If, on the other hand, we use Minitab for Windows to carry out a hypothesis test, steps 5 and 6 will be slightly different, as follows:

Step 5. Find the 'p value' on the computer output.

Step 6. Compare the 'p value' with the significance level, and decide whether to reject the null hypothesis, H_c.

The 'p value' method will be discussed in detail in Section 10.9.

What we can do if we decide that the assumptions are not valid (step 7 above) is discussed later in this chapter in Section 10.17.

In the following sections, each of the seven steps and the underlying concepts will be explained, with the aid of a simple example.

10.2 What is a Hypothesis?

In terms of the examples of the previous chapter, a hypothesis is a statement about the value of a population parameter, such as the population mean, μ. We use the sample data to decide whether the stated value of the parameter is reasonable. If we decide that it is not reasonable we reject the hypothesis in favour of another hypothesis. It is important to note at this stage, then, that in hypothesis testing we have two hypotheses to consider. Using sample data, we decide which hypothesis is the more reasonable. We call the two hypotheses the **null hypothesis** and the **alternative hypothesis**.

10.3 Which is the Null Hypothesis and Which is the Alternative Hypothesis?

The null hypothesis generally expresses the idea of 'no difference'—think of 'null' as meaning 'no'. In terms of the examples of the previous chapter a null hypothesis could be a statement that the mean of a population is 'no different from', that is 'equal to', a specified value. The notation we will use to denote a null hypothesis is H_0.

The null hypothesis

$$H_0 : \mu = 165$$

states that the population mean equals 165.

The alternative hypothesis, which we denote by H_1, expresses the idea of 'some difference'. Alternative hypotheses may be **one-sided** or **two-sided**. The first two examples below are one-sided since each specifies only one side of the number 165; the third example is two-sided since both sides of the number 165 are specified:

$H_1 : \mu > 165$	(Population mean greater than 165)
$H_1 : \mu < 165$	(Population mean less than 165)
$H_1 : \mu \neq 165$	(Population mean not equal to 165)

In each hypothesis test we perform, we should specify both the null and the alternative hypotheses appropriate to the purpose of our study or investigation, and **before the sample data are collected**. Remember that we use our sample data to test the null hypothesis, and not the other way round. Then, if we reject the null hypothesis we should accept the alternative hypothesis; while if we do not reject the null hypothesis, we should reject the alternative hypothesis. It is WRONG to use your sample data to suggest a null hypothesis and then to test this hypothesis using the same sample data.

10.4 What is a Significance Level?

Hypothesis testing is also sometimes referred to as **significance testing**. The concept of **significance level** is similar to the concept of confidence level. The usual value we choose for our significance level is 5%, just as we usually choose a confidence level of 95%. Just as the confidence level expresses the idea that we would be prepared to bet heavily that the interval we state actually does contain the value of the population parameter of interest, so a significance level of 5% expresses a similar idea in connection with hypothesis testing. For example, a significance level of 5% is the risk we take in rejecting the null hypothesis, H_0, in favour of the alternative hypothesis, H_1, when in reality H_0 is the correct hypothesis.

Example

If the first three steps of our hypothesis test are

1. $H_0 : \mu = 165$
2. $H_1 : \mu \neq 165$
3. 5% significance level,

then we are stating we are prepared to run a 5% risk that we will reject H_0 and conclude that the mean is not equal to 165, when the mean is actually equal to 165.

We cannot avoid the small risk of drawing such a wrong conclusion in hypothesis testing because we are trying to draw conclusions about a population using only part of the information in the population, namely, the sample data. The corresponding risk in confidence interval estimation is the small risk we take that the interval we calculate will **not** contain the true value of the population parameter of interest. For example, in calculating a 95% confidence interval for the population mean, there is a 5% risk that this interval will not contain the true value of μ.

10.5 What is a Test Statistic, and How do We Calculate It?

A test statistic is a value we can calculate from our sample data and from the value of the parameter we specify in the null hypothesis, using an appropriate formula.

Example

If the first three steps of our hypothesis test are as in the example of Section 10.4, and our sample data are summarised as

$$\bar{x} = 162.3, \quad s = 5.3, \quad n = 9.$$

then the fourth step of our hypothesis test is as follows:

4. *Calc t* is obtained using the formula

$$Calc\ t = \frac{\bar{x} - \mu}{\frac{s}{\sqrt{n}}} \qquad (10.1)$$

$$= \frac{162.3 - 165}{\frac{5.3}{\sqrt{9}}}$$

$$= -1.53.$$

where μ refers to the value stated in the null hypothesis.

10.6   How do We Find the Tabulated Test Statistic?

We must know which tables to use for a particular application, and (of course) how to use them.

Example

1. $H_0 : \mu = 165$
2. $H_0 : \mu \neq 165$
3. 5% significance level
4. *Calc t* $= -1.53$, from above and assuming the same data summary, i.e., $\bar{x} = 162.3$, $s = 5.3$, $n = 9$.
5. The appropriate table is Table C.5, and we enter the tables for:
    (a) $\alpha = 0.05/2$, dividing by 2 since H_1 is two-sided. So, $\alpha = 0.025$.
    (b) $\nu = (n-1) = 9-1 = 8$ degrees of freedom.

So, the tabulated test statistic is *Tab t* $= 2.306$ from Table C.5.

10.7   How do We Compare the Calculated and the Tabulated Test Statistics?

For the example in Section 10.6 we reject H_0 if $|Calc\ t| > Tab\ t$, where the vertical lines mean that we ignore the sign of *Calc t* and consider only its magnitude (e.g., $|-5| = 5$, $|5| = 5$).

Since, in this example, $|Calc\ t| = 1.53$, and *Tab t* $= 2.306$, we do not reject H_0. Fig. 10.2 shows that only calculated values of t in the 'tails' of the distribution, beyond the **critical values** of -2.306 and $+2.306$, lead to the rejection of H_0.

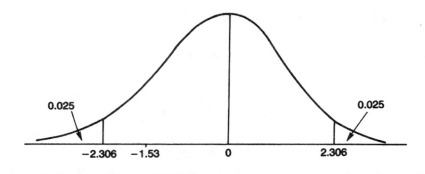

Figure 10.2 *t* **Distribution for** $\nu = 8$ **Degrees of Freedom**

10.8 What is Our Conclusion, and What Assumptions Have We Made?

Our conclusion should be a sentence in words, as far as possible devoid of statistical terminology. For the example used previously in this chapter, since we decided not to reject H_0 in favour of an alternative stating that the mean differed from 165, we conclude that 'the mean is not significantly different from 165, at the 5% level of significance'.

The only assumption of this test is that the variable is approximately normally distributed, which we have already seen (in the example in Section 9.4) is a reasonable assumption to make in this case.

10.9 Using p Values Instead of Tables

When we use Minitab for Windows to perform a hypothesis test, it is not necessary to use Statistical Tables, such as Table C.5. Minitab will calculate the value of *Calc t*, for example, as in Section 10.5 where Formula (10.1) was used. Minitab then calculates a 'p value' corresponding to the calculated value of the test statistic, e.g., *Calc t*, and taking into account whether H_1 is two-sided or one-sided.

As far as the seven-step method is concerned, when we use the 'p value' method only steps 5 and 6 are different from those listed in Section 10.1. Calling the two new steps 5a and 6a, respectively, in general terms:

Step 5a. Find the p value on your computer output.

Step 6a. Assuming a significance level of 5%, reject H_0 if the p value is less than 0.05. If, on the other hand, p is greater than or equal to 0.05, H_0 is not rejected.

Note: The 'p value method' (as opposed to the 'Tables method') should only be used if you have a computer package, such as Minitab. Do not attempt to calculate p values using a basic hand-held calculator!

10.10 Hypothesis Test for the Mean, μ, of a Population

In this section we summarise the seven steps for the example used earlier in this chapter, using the 'Tables method'. We also show how to perform the same test example using Minitab for Windows, and how to interpret the computer output.

1. $H_0: \mu = 165$
2. $H_1: \mu \neq 165$
3. 5% significance level
4. The calculated test statistic is

$$Calc \ t = \frac{\bar{x} - 165}{\frac{s}{\sqrt{n}}} = \frac{163.4 - 165}{\frac{5.3}{\sqrt{9}}} = -1.53$$

5. *Tab t* = 2.306, for $\alpha = 0.025$ and $\nu = 9-1 = 8$.
6. Since $|Calc \ t| < Tab \ t$, do not reject H_0.
7. The mean is not significantly different from 165 (5% significance level)

Assumption: Variable is approximately normally distributed.

Notice that although we did not reject H_0, neither did we conclude that $\mu = 165$. We cannot be so definite, given that we have only a sample from the whole population, and we recall that μ is the mean of the population. The conclusion in step 7 simply implies that we think that H_0 is a more reasonable hypothesis than H_1 in this example. Clearly, we cannot conclude that $\mu = 165$ and $\mu = 164$ and so on.

Using Minitab for Windows

Enter the 9 values for the heights of 9 students (from Section 9.4) into C1, say. Then:
    Choose **Stat > Basic Stats > 1-Sample *t***
    Enter **C1** in **Variable** box
    Choose **Test Mean**
    Enter **165** in **Test Mean** box
    Choose **'not equal'** in **Alternative** box
    Click on **OK**

The Minitab output is as follows:

TEST OF MU = 165.00 VS MU N.E. 165.00

	N	MEAN	STDEV	T	P VALUE
C1	9	162.33	5.34	−1.50	0.17

Notes: Step 5a is 'p value' = 0.17.

Step 6a is 'Do not reject H_0, since $0.17 > 0.05$'.

10.11 Two Examples of Tests with One-Sided Alternative Hypotheses

If we had chosen a one-sided H_1 in the previous example, the steps would have varied a little. Since we could have chosen $\mu > 165$ or $\mu < 165$ as our alternative, both of these cases are now given below, side by side, using the 'Tables method'. These should be read with reference to Fig. 10.3. Then the 'p value method' is shown, once again using Minitab for Windows.

1.	$H_0 : \mu = 165$	$H_0 : \mu = 165$
2.	$H_1 : \mu > 165$	$H_0 : \mu < 165$
3.	5% significance level	5% significance level
4.	Calc t = −1.53	Calc t = −1.53
5.	Tab t = 1.860	Tab t = 1.860
	for $\alpha = 0.05/1 = 0.05$,	for $\alpha = 0.05/1 = 0.05$,
	and $\nu = (n − 1) = 8$	and $\nu = (n − 1) = 8$
	Since Calc t < Tab t, do not reject H_0	Since Calc t > − Tab t, do not reject H_0
7.	The mean is not significantly greater than 165 (5% level)	The mean is not significantly less than 165 (5% level)

Assumption

Variable is approximately normally distributed.

Fig. 10.3 shows that only calculated values of t in the right-hand tail, greater than the critical value of 1.860, lead to rejection of H_0.

Variable is approximately normally distributed.

Fig. 10.3 shows that only calculated values of t in the left-hand tail, less than the critical value of −1.860, lead to rejection of H_0.

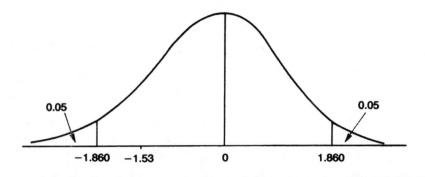

Figure 10.3 *t* Distribution for $\nu = 8$ Degrees of Freedom

The six lines of Minitab for Windows instructions given earlier in this section for the two one-sided cases are the same, except that 'not equal' in the fifth line is replaced by 'greater than' or 'less than' in turn. The computer output is very similar, except that 'N.E.' is replaced by 'G.T.' or 'L.T.', respectively, while the p values are 0.086 and 0.91, respectively. Neither p value is less than 0.05, so H_0 is not rejected in either case.

Note that tests in which the alternative hypothesis is two-sided are often referred to (in other texts) as **two-tailed** tests, while tests in which the alternative hypothesis is one-sided are often referred to as **one-tailed** tests.

10.12   Hypothesis Test for a Binomial Probability

Suppose we wish to test a hypothesis for p, the probability of success in a single (Bernoulli) trial, using sample data from a number of such trials. If x successes resulted from n trials, the test statistic is calculated using Formula (10.2):

$$Calc\ z = \frac{\frac{x}{n} - p}{\sqrt{\frac{p(1-p)}{n}}} \tag{10.2}$$

where p is the value specified in the null hypothesis.
We can use this formula if $np > 5$, and $n(1-p) > 5$. The tabulated test statistic is *Tab z*, obtained from Table C.3(b).

Example

Test the hypothesis that the percentage of voters who will vote for party A in an election is 50% against the alternative that it is greater than 50%, using the random sample data from an opinion poll that 110 out of 200 voters said they would vote for party A.

1. $H_0: p = 0.5$, which implies that 50 % of the population will vote for party A.
2. $H_1: p > 0.5$, which implies that party A will have an overall majority.
3. 5% significance level.
4. The test statistic formula here is Formula (10.2). Since $np = 200 \times 0.5 = 100$ and $np(1-p) = 200 \times 0.5 \times 0.5 = 50$ are both greater than 5, the two conditions for using this formula are both satisfied

$$Calc\ z = \frac{\frac{110}{200} - 0.5}{\sqrt{\frac{0.5(1 - 0.5)}{200}}} = 1.414$$

5. *Tab* $z = 1.645$, since in Table C.3(b) this value of z corresponds to a tail of 0.05/1, the significance level divided by 1 since H_1, is one sided (see Fig. 10.4).
6. Since *Calc z* < *Tab z*, do not reject H_0.
7. The percentage of voters for party A is not significantly greater than 50%, at the 5% significance level. So it is not reasonable to conclude that party A will gain an overall majority in the election.

Assumptions: (a) The four binomial conditions apply (see Section 6.3).
(b) Voters tell the truth and don't change their minds; see the same data in Section 9.8, where we calculated a 95% C.I. for *p*.

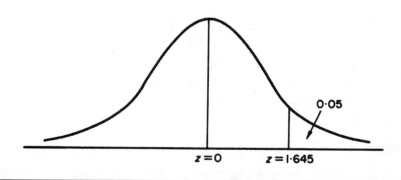

Figure 10.4 Standardized Normal Distribution

10.13 Hypothesis Test for the Mean of a Population of Differences, 'Paired' Samples Data

Example

For the example given in Section 9.10 in which two methods of teaching children to read were compared, suppose we want to decide whether the new method, N, is better than the standard method, S, in terms of the mean difference in the test scores of the two methods. We assume that we have the same data as in Table 9.2, and hence the same summary of those data:

$$\bar{d} = 7.5, \quad s_d = 8.48, \quad n = 10.$$

1. $H_0 : \mu_d = 0$. This implies that the mean of the population of differences is zero, in other words that the two teaching methods give the same mean test score.
2. $H_1 : \mu_d > 0$. Since the differences were calculated using (N score $-$S score), this implies that the N method gives a higher mean test score than the S method.
3. 5% significance level.
4. For this 'paired samples t test', the test statistic formula is Formula (10.3)

$$Calc\ t = \frac{\bar{d}}{\frac{s_d}{\sqrt{n}}} = \frac{7.5}{\frac{8.48}{\sqrt{10}}} = 2.80. \tag{10.3}$$

5. $Tab\ t = 1.833$, for $\alpha = 0.05$, since H_1 is one-sided, and $\nu = (n-1) = 9\ d.f.$
6. Since $Calc\ t > Tab\ t$, reject H_0.
7. The N method gives a significantly higher mean test score than the S method (5% level).

Assumption: The differences must be approximately normally distributed, and we saw in Section 9.10 that this was a reasonable assumption to make for these data.

In order to use Minitab for Windows to perform a 'paired samples t test', simply follow the method of the example in Section 10.10, by first entering the 10 differences into C1, say:

Choose **Stat** > **Basic Stats** > **1-Sample** t

Enter **C1** in the **Variables** box
Choose **Test Mean**
Enter **0** in **Test Mean** box
Choose **'greater than'** in **Alternatives** box
Click on **OK**

The Minitab output is as follows:

	N	MEAN	STDEV	T	P
TEST OF MU = 0 VS MU G.T. 0					
C1	10	7.5	8.48	2.80	0.010

Since the 'p value' is less than 0.05, we reject H_0 and conclude that the mean of the population of differences is greater than zero, i.e., that the N method gives a significantly higher mean score than the S method. This is the same conclusion as that obtained earlier in this section using the 'tables method'.

N.B. The 95% confidence interval for μ_d was 1.4 to 13.6 (see Section 9.10), which also implies the same conclusion as above, since both limits are positive, and recalling that the differences were calculated using: difference = N score − S score.

10.14 Hypothesis Test for the Difference Between the Means of Two Populations, 'Unpaired' Samples Data

Example

For the example given in Section 9.11, comparing the A-level counts of BA and BSc students, suppose we want to test whether the mean A-level counts of the two populations are equal. We assume that we have the same data as before, which were summarised as follows:

$$\bar{x}_1 = 16.22, \quad s_1 = 7.10, \quad n_1 = 9$$
$$\bar{x}_2 = 10.72, \quad s_2 = 4.78, \quad n_2 = 31$$

The seven steps of 'unpaired samples *t* test' are as follows:
1. $H_0: \mu_1 = \mu_2$. This implies there is no difference between the mean A-level counts of the two populations.
2. $H_1: \mu_1 \neq \mu_2$. This implies that there is a difference, in one direction or the other.

3. 5% significance level.
4. For this 'unpaired samples t test', the formula for the calculated test statistic is Formula (10.4), but first we need to obtain s using Formula (9.7). However we have already done this calculation in Section 9.11! We found that $s = 5.35$. Hence:

$$Calc \ t = \frac{16.22 - 10.71}{5.35\sqrt{\frac{1}{9} + \frac{1}{31}}} = \frac{5.51}{2.03} = 2.72$$

5. $Tab \ t = 2.02$, for $\alpha = 0.05/2 = 0.025$ and $\nu = 9 + 31 - 2 = 38$ d.f.
6. Since $Calc > Tab \ t$, i.e., $2.72 > 2.02$, reject H_0.
7. The mean A-level counts for BA and BSc students are significantly different (5% level of significance). The direction of the difference is clear since $\bar{x}_1 > \bar{x}_2$. Hence we can conclude that the mean A-level count for BA students is significantly higher than for BSc students. In fact, we have already drawn this conclusion in Section 9.11 by considering only the 95% confidence interval for $(\mu_1 - \mu_2)$. We discuss the connection between the topics of Chapters 9 and 10 in Section 10.18 below.

Assumptions: First, the measurements in each population must be approximately normally distributed, this assumption being less critical the larger the values of n_1 and n_2. Second, the population standard deviations, σ_1 and σ_2, must be equal. For these data, both assumptions are reasonable, as discussed in Section 9.11, since the same assumptions apply to the caculation of confidence intervals for $(\mu_1 - \mu_2)$.

In order to carry out an 'unpaired samples t test' using Minitab for Windows, we will use the A-level data used already in this section. We will assume that the 9 A-level counts for BA students have been placed into C1 of a Minitab spreadsheet, while the 31 A-level counts for BSc students have been placed in C2.

Then the following steps will make Minitab carry out a t test and produce an output as shown below:

Choose **Stat > Basic Stats > 2-sample t**
Choose **samples in different columns**
Enter **C1** in **first** box
Enter **C2** in **second** box
Choose **not equal to** in **Alternative** box
Choose **Assume equal variances** (click in little box)
Click on **OK**
The output is as follows:

TWO SAMPLE T FOR C1 VS C2

	N	MEAN	STDEV	SE MEAN
C6	9	16.22	7.10	2.4
C7	31	10.71	4.78	0.86

95 PCT C1 FOR MU C1 − MU C2: (1.4, 9.61)
TTEST MU C1 = MU C2 (VS NE) : T=2.72 P=0.0098 DF=38
POOLED STDEV = 5.35

Since the p value is 0.0098, i.e., less than 0.05, H_0 is rejected. This, of course, agrees with the conclusion using the 'tables method' (see earlier in this section).

10.15 Hypothesis Test for the Equality of the Variances of Two Normally Distributed Populations

In Section 9.11 we obtained a confidence interval for $(\mu_1 - \mu_2)$, the difference between the means of two populations. One of the assumptions needed to use the correct Formula (9.6) was that the two population standard deviations were equal. This is the same assumption we have just needed for the unpaired samples t test in Section 10.14. In Section 9.11 we could not easily decide whether this assumption was valid, since the only method we knew at that time was to look at the values of s_1 and s_2, the sample estimates of σ_1 and σ_2.

In this section we will carry out a formal test of the equality of the two variances. This is known as the 'F test'.

Example

For the sample data in Section 9.11, we note that:

$$s_1 = 7.10, \quad n_1 = 9, \quad s_2 = 4.78, \quad n_2 = 31$$

1. $H_0 : \sigma_1^2 = \sigma_2^2$
2. $H_1 : \sigma_1^2 \neq \sigma_2^2$
3. 5% significance level
4. Since $s_1 > s_2$, Calc $F = \frac{s_1^2}{s_2^2}$ (if $s_1 < s_2$, Calc $F = \frac{s_2^2}{s_1^2}$). Therefore, Calc $F = \frac{7.10^2}{4.78^2} = 2.21$.
5. Tab $F = 2.27$, using Table C.6 for $\nu_1 = 9 - 1 = 8$, $\nu_2 = 31 - 1 = 30$, where ν_1 is the number of d.f. associated with the numerator of Calc F, i.e., 7.10. Similarly, ν_2 is the number of d.f. associated

with the denominator of *Calc F,* i.e., 4.78. In Table C.6, we look up ν_1 along the top of the table and ν_2 down the side of the table.

6. Since *Calc F* < *Tab F, H_0* is not rejected.
7. Hence the variances are not significantly different (5% level).

Assumptions: Both populations are normally distributed. We were able to justify this using dotplots in Section 9.11. However, if you require a less subjective test of normality than that afforded by dotplots, please refer to Section 16.5.

10.16   The Effect of Choosing Significance Levels Other Than 5%

Why do we not choose a significance level lower than 5%, since we would run a smaller risk of rejecting H_0 when H_0 is correct (refer to Section 10.4 if necessary). Just as there are advantages and disadvantages in choosing a confidence level above 95% — a consequence of a higher confidence level is a wider confidence interval — a similar argument applies to significance levels below 5%.

If we reduce the significance level to below 5%, we reduce the risk of wrongly rejecting H_0, but we increase the risk of drawing a different wrong conclusion, namely, the risk of wrongly rejecting H_1. Nor can we set both risks at 5% for the examples described in this chapter (for reasons which are beyond the scope of this book — interested readers will find a discussion in Chapter 14 of *Essential Statistics for Medical Practice,* by D.G. Rees, Chapman & Hall, 1994, of this and related topics). Even if we could set both risks at 5% it might not be a wise thing to do! Consider the risks in a legal example and judge(!) whether *they* should be equal:

(a)   The risk of convicting an innocent man in a murder trial.
(b)   The risk of releasing a guilty man in a murder trial.

There is nothing sacred about the '5%' for a significance level, nor the '95%' for a confidence level, but we should be aware of the consequences of departing from these conventional levels.

10.17   What if the Assumptions of a Hypothesis Test are not Valid?

If at least one of the assumptions of a hypothesis test is not valid, i.e., there is insufficient evidence to make us believe that they are all reasonable assumptions, then the test is also invalid and the conclusions may well be wrong.

In such cases, alternative tests, called distribution-free tests, or more commonly, **nonparametric** tests, should be used if they are available. These tests do not require such rigorous assumptions as the 'parametric' tests described earlier in this chapter, but they have the disadvantage that they are less **powerful,** meaning that we are less likely to accept the alternative hypothesis (as a consequence of rejecting the null hypothesis), when the alternative hypothesis is correct. Some nonparametric tests are described in Chapter 11.

10.18 The Connection Between Confidence Interval Estimation and Hypothesis Testing

Confidence interval estimation and hypothesis testing provide similar types of information. However, a confidence interval (if a formula exists to calculate it) provides more information than the corresponding hypothesis test.

Example

Consider the student height data in the last example of Section 9.4. For these data we know that: $\bar{x} = 162.4$, $s = 5.3$, $n = 9$. We also calculated a 95% confidence interval for the population mean, μ, to be 158.2 to 166.4 cm.

From this result we can immediately state that any null hypothesis specifying a value of μ within this interval would not be rejected in favour of the two-sided alternative hypothesis, assuming a 5% level of significance. So, for example, $H_0: \mu = 160$ would not be rejected in favour of $H_1 : \mu \neq 160$, but $H_0: \mu = 150$ would be rejected in favour of $H_1 : \mu \neq 150$, and $H_0: \mu = 170$ would be rejected in favour of $H_1: \mu \neq 170$.

Generalizing on the example above, we can state that:

> **A confidence interval for a population parameter contains the range of values for the parameter we would not wish to reject.**

The confidence interval is a way of representing all the null hypotheses we would not wish to reject, on the evidence of the sample data. To this extent, a confidence interval contains much more information than the conclusion of a hypothesis test.

10.19 Summary

A statistical hypothesis is often a statement about the value of a population parameter. In a seven-step method we use sample data to decide whether to reject the null hypothesis in favour of an alternative hypothesis. If the assumptions of a test are not valid, alternative nonparametric tests (to be discussed in Chapter 11) may be available.

The connection between confidence interval estimation and hypothesis testing was discussed; the former contains more information than the latter.

Worksheet 10: Hypothesis Testing Including t, z, and F Tests

1. What is
   (a) A (statistical) hypothesis?
   (b) A null hypothesis?
   (c) An alternative hypothesis?
   Give an example of each in a subject area of your choice.
2. What is a significance level?
3. Why do we need to run a risk of wrongly rejecting the null hypothesis?
4. Why do we choose 5% as the risk of wrongly rejecting the null hypothesis?
5. How can we tell whether an alternative hypothesis is one-sided or two-sided?
6. How do we know whether to specify a one-sided or a two-sided alternative hypothesis in a particular investigation? Think of an example when each would be appropriate.

Questions 7 to 11 inclusive are multiple choice. Choose one of the three options in each case.

7. The significance level is the risk of:
   (a) Rejecting H_0 when H_0 is correct,
   (b) Rejecting H_0 when H_1 is correct,
   (c) Rejecting H_1 when H_1 is correct.
8. If we decide not to reject a null hypothesis H_0 this:
   (a) Proves that H_0 is true,
   (b) Proves that H_1 is false,
   (c) Implies that H_0 is likely to be true.
9. If the magnitude of the calculated value of t is less than the tabulated value of t, and H_1 is two-sided, we should:
   (a) Reject H_0,
   (b) Not reject H_0,
   (c) Accept H_1.
10. The t test for samples from a normal population must be used when:
    (a) The sample size is small,

(b) The standard deviation is unknown,

(c) The sample is small and the standard deviation is unknown.

11. In an unpaired samples *t* test with sample sizes of 10 and 10, the value of tabulated *t* should be obtained for:

(a) 9 degrees of freedom,

(b) 19 degrees of freedom,

(c) 18 degrees of freedom.

In Questions 12 to 23 inclusive, use a 5% significance level unless otherwise stated. In each question the assumptions required for the test should be stated, and you should also decide whether the assumptions are likely to be valid.

12. Eleven cartons of sugar, each nominally containing 1 kg, were randomly selected from a large batch of cartons. The weights of sugar were 1.02, 1.05, 1.08, 1.03, 1.00, 1.06, 1.08, 1.01, 1.04, 1.07, and 1.00 kg. Do these data support the hypothesis that the mean weight for the batch is 1 kg?

13. A cigarette manufacturer claims that the mean nicotine content of a brand of cigarettes is 0.30 mg per cigarette. An independent consumer group selected a random sample of 1000 cigarettes and found that the sample mean was 0.31 mg per cigarette, with a standard deviation of 0.03 mg. Is the manufacturer's claim justified or is the mean nicotine content significantly higher than he states?

14. The weekly take-home pay (£) of a random sample of 30 farm workers is as folllows:

85	85	95	95	95	95	95	105	105	105
105	105	105	105	105	105	115	115	115	115
115	115	115	115	115	115	115	125	125	125

Do these data support the claim that the mean weekly take-home pay of all farm workers is below £110?

15. The market share for the E10 size packet of a particular brand of washing powder has averaged 30% for a long period. Following a special advertising campaign it was discovered that, of a random sample of 100 people who had recently bought the E10 size packet, 35 had bought the brand in question. Has its market share increased?

16. Of a random sample of 300 gourmets, 180 prefer thin soup to thick soup. Is it reasonable to expect that 50% of all gourmets prefer thin soup?

17. It is known from the records of an insurance company that 14% of all males holding a certain type of life insurance policy die during their 60th, 61st, or 62nd year. The records also show that, of

a randomly selected group of 1000 civil servants holding this type of policy, 112 died during their 60th, 61st, or 62nd year. Is it reasonable to conclude that male civil servants have a lower death rate than 14% during these years?

18. An experiment was conducted to compare the performance of two varieties of wheat, A and B. Seven farms were randomly chosen for the experiment, and the yields (in tonnes per hectare) for each variety on each farm were as follows:

	Farm Number						
	1	2	3	4	5	6	7
Yield of variety A	4.6	4.8	3.2	4.7	4.3	3.7	3.1
Yield of variety B	4.1	4.0	3.5	4.1	4.5	3.2	3.8

(a) Why do you think both varieties were tested on each farm, rather than testing variety A on seven farms and variety B on seven other farms?

(b) Carry out a hypothesis test to decide whether the mean yields are the same for the two varieties.

19. A sample length of material was cut from each of five randomly selected rolls of cloth and each length was divided into two halves. One half was dyed with a newly developed dye, and the other half with a dye that had been in use for some time. The ten pieces were then washed and the amount of dye washed out was recorded for each piece as follows:

	Roll				
	1	2	3	4	5
Old dye	13.2	13.7	15.4	13.5	16.8
New dye	12.5	14.3	16.8	14.9	17.4

Investigate the allegation that the amount of dye washed out for the old dye is significantly less than for the new dye.

20. A sleeping drug and a neutral control were tested in turn on a random sample of ten patients in a hospital. The data below represent the differences between the number of hours of sleep under the drug and the neutral control for each patient:

2.0   0.2   − 0.4   0.3   0.7   1.2   0.6   1.8   − 0.2   1.0

Is it reasonable to conclude that the drug would give more hours of sleep on average than the control for all the patients in the hospital?

21. To compare the strengths of two cements, six mortar cubes were made with each cement and the following were recorded:

Cement A	4600	4710	4820	4670	4760	4480
Cement B	4400	4450	4700	4400	4170	4100

Is there a significant difference between the mean strengths of the two cements? In answering this question, state two assumptions you need to make, and test one of them. (Hint: use an F test first.)

22. The price of a standard 'basket' of household goods was recorded for 25 corner shops selected at random from the suburbs of all the cities in England. In addition, 25 food supermarkets were also randomly selected from the main shopping centres of the cities. The data were summarized as follows:

	Corner Shop	Supermarket
Mean price (£)	19.45	18.27
Standard deviation of price (£)	1.20	1.00
Number in sample	25	25

Test the hypothesis that corner shops are charging the same on average as supermarkets for the standard basket.

23. Two geological sites were compared for the amount of vanadium (parts per million) found in clay. It was thought that the amount would be less for area A than for area B. Do the following data, which consist of the results from 10 samples taken randomly from each area, support this view?

Area A	Area B
50	50
65	70
75	90
80	95
90	95
95	100
105	105
110	110
130	125
140	150

24. Use the 95% confidence interval method on the data in Questions 12, 16, 18, and 21 to check the conclusions you reached using the hypothesis testing method.

25. Referring to Question 10 of Worksheet 8 concerning a project to decide whether students who undertake paid employment during term-time perform academically as well as similar students who did not take on paid employment during term:
    (a) Can you see a way to answer part (f) now that you have (presumably) covered Chapters 9 and 10?
    (b) What might you conclude now about the differences in academic performance between the two types of student?
    (c) What reservations might you have concerning your conclusions?
    (d) Might it be possible to conclude that paid term-time employment has a detrimental effect on academic performance?
    Note: Look at this Question again when you have covered Chapter 11 and Worksheet 11.

Chapter 11

Nonparametric Hypothesis Tests

At least the difference is very inconsiderable.

11.1   Introduction

There are some hypothesis tests which do not require such rigorous assumptions as the tests described in Chapter 10. However, these **'nonparametric' tests are less powerful** than the corresponding 'parametric' tests, which means that we are less likely to reject the null hypothesis, and hence accept the alternative hypothesis, when the alternative hypothesis is, in fact, correct. We can therefore conclude that parametric tests are generally preferred if the assumptions of these tests can be shown to be valid.

Three nonparametric tests are described in this chapter for which there are roughly equivalent t tests, as indicated in Table 11.1.

11.2   Sign Test for the Median of a Population

Example

Suppose we collect the weekly incomes (£) of a random sample of 10 self-employed window cleaners:

200   550   290   170   180   350   190   210   160   250

Table 11.1 Nonparametric Tests

Nonparametric Test	Application	Reference for Roughly Equivalent t Test
Sign test	Median of a population	Section 10.10
Sign test	Median of a population of differences — 'paired' samples data	Section 10.13
Wilcoxon signed rank test	Median of a population of differences — 'paired' samples data	Section 10.13
Mann-Whitney U test	Difference between the medians of two populations — 'unpaired' samples data	Section 10.14

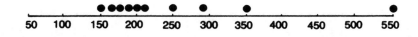

Figure 11.1 Dotplot of the Incomes of the Ten Self-Employed Window Cleaners

Let's suppose that we are interested in testing a hypothesis concerning some average value of the weekly incomes of self-employed window cleaners.

A dotplot (see Fig. 11.1) indicates positive skewness, and in fact the measure of skewness is 1.27 using Formula (4.8) (see Section 4.13).

Since the measure of skewness is greater than $+1$, the assumption of normality is not reasonable, and so a t test would not be valid here.

Instead, we will carry out a test that the **median** is £250 per week. This can be done by carrying out a **sign test** as follows.

The sign test for the median of a population requires that we write down the signs of the differences between each income in our sample and the hypothesised value of £250. Using the convention that incomes above £250 correspond to differences that are positive $(+)$, incomes below £250 correspond to differences that are negative $(-)$, while incomes of exactly £250 (which are referred to as *ties*) are ignored, the signs for the above data are

$$- \quad + \quad + \quad - \quad - \quad + \quad - \quad - \quad -$$

There are three $+$ signs, six $-$ signs and one tie. If the null hypothesis (that the median is £250) is true, the probability of an income above (or below)

£250 is 0.5. We will take the alternative hypothesis to be that the median is not equal to be £250. The seven-step method for the sign test for this example is

1. $H_0: p(+) = p(-) = 0.5$, where $p(+)$ and $p(-)$ mean the probability of a + sign and a − sign, respectively.
2. $H_1: p(+) \neq p(-)$, a two-sided alternative hypothesis.
3. 5% significance level.
4. For the sign test, the calculated test statistic is a binomial probability of getting the result obtained *or a result which is more extreme*, assuming for the moment and for the purposes of the calculation only, that H_0 is true. For this example we need to calculate:

    P(6 or more minus signs in 9 trials when $p(-) = 0.5$)

    $= P(6) + P(7) + P(8) + P(9)$

    $= \binom{9}{6}0.5^6 0.5^3 + \binom{9}{7}0.5^7 0.5^2 + \binom{9}{8}0.5^8 0.5^1 + \binom{9}{9}0.5^9 0.5^0$

    $= (84 + 36 + 9 + 1)(0.5)^9$

    $= 0.2539.$

5. The tabulated test statistic for the sign test is simply the significance level divided by 2, if H_1 is two-sided, and so equals 0.025 for this example.
6. Reject H_0 if the calculated probability is less than (significance level)/2, for a two-sided alternative. For this example, since $0.2539 > 0.025$, we do not reject H_0.
7. The median wage is not significantly different from £250 (5% level).

Assumption: The variable, here income, has a continuous distribution.

Notes

(a) Instead of P(6 or more minus signs in 9 trials when $p(-) = 0.5$), we could have calculated P(3 or fewer plus signs in 9 trials when $p(+) = 0.5$), but the answer would have been the same, because of the symmetry of the binomial distribution when $p = 0.5$.
(b) Notice that the assumption required of a continuous distribution is less restrictive than the assumption of a normal distribution.
(c) If $n > 10$, we can alternatively use the method of Section 11.4.

(d) Since we did not use the magnitudes of the differences (between each income and £250), this test can be performed even if we do not know the actual sizes of the differences, but simply their signs.

11.3 Sign Test for the Median of a Population of Differences, 'Paired' Samples Data

Example

For the example given in Section 9.10 concerning two methods of teaching children to read, suppose we want to decide whether the new method (N) is better than the standard method (S), but we do not wish to assume that the differences in the test scores are normally distributed. Instead, we can use the sign test to decide whether the median score by the new method is significantly greater than that by the standard method.

The differences (N score − S score) were

$$7 \quad -2 \quad 6 \quad 4 \quad 22 \quad 15 \quad -5 \quad 1 \quad 12 \quad 15.$$

1. $H_0: p(+) = p(-) = 0.5$. The median of the population of differences is zero, which implies that the two methods are equally effective.
2. $H_1: p(+) > p(-)$. Median of N scores is greater than the median of the S scores, which implies that method N is more effective than method S.
3. 5% significance level.

4 and 5. If the null hypothesis is true we would expect equal numbers of + and − signs. If the alternative hypothesis is true we would expect more + signs, so the null hypothesis is rejected if:

P(observed number *or more* of + signs out of 10 when $p(+) = 0.5$) <0.05, for a one-sided alternative hypothesis.

In the example we have 8 + signs in 10 differences.

$$P(8 \text{ or more}) = 1 - P(7 \text{ or fewer})$$
$$= 1 - 0.9453$$
$$= 0.0547, \text{ using Table C.1 (Appendix C)}$$

6. Since $0.0547 > 0.05$ (if only just), we do not reject the null hypothesis.
7. The median score by the N method is not significantly greater than that by the S method (5% level).

Assumption: The differences are continuous, which can be assumed for test scores even if we quote them to the nearest whole number.

Notes

(a) Differences of zero are ignored in this test.
(b) If $n > 10$, we can alternatively use the method of Section 11.4.
(c) As in Section 11.3, this test can be used in cases where we do not know or cannot quantify the magnitudes of the differences, for example in preference testing. Since only the signs of the differences are used in the test, we could adopt the convention:
   'brand A preferred to brand B' recorded as +, and
   'brand B preferred to brand A' recorded as −, and
   'no preference' cases are ignored, as they are the equivalent of differences of zero.

11.4   Sign Test for Large Sample Sizes ($n > 10$)

The sign test for sample sizes larger than 10 is made easier by the use of a normal approximation method (similar to that used in Section 7.7) by putting:

$$\mu = \frac{n}{2} \quad \text{and} \quad \sigma = \frac{\sqrt{n}}{2} \tag{11.1}$$

Example

Suppose that for $n = 30$ paired samples there are 20+ and 10− differences.

1. $H_0: p(+) = p(-) = 0.5$. The median of the population of differences is zero.
2. $H_1: p(+) \neq p(-)$ (two-sided).
3. 5% significance level.
4. Following the method used in the example in Section 11.2, we now need to calculate P(20 or more + signs in 30 trials, when $p(+) = 0.5$).
   Instead of calculating several binomial probabilities, we can apply Formula (11.1):

$$\mu = \frac{30}{2} = 15 \quad \text{and} \quad \sigma = \frac{\sqrt{30}}{2} = 2.74$$

From now on, we will use the normal distribution with these parameters (see Fig. 11.2 below).

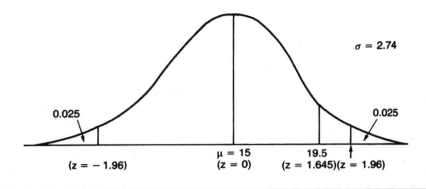

Figure 11.2   A Normal Distribution With μ = 15, σ = 2.74

We also need to introduce a continuity correction since '20 or more on a discrete scale' is equivalent to 'more than 19.5 on a continuous scale' (recall that the binomial is a discrete distribution, while the normal is a continuous distribution). For $x = 19.5$,

$$Calc \ z = \frac{(x - \mu)}{\sigma} = \frac{(19.5 - 15)}{2.74} = 1.64.$$

5. *Tab z* $= 1.96$ from Table C.3(b), since this value of z corresponds to a tail area of $0.05/2$, the significance level divided by 2 since the alternative hypothesis is two-sided.
6. Since $|Calc \ z| < Tab \ z$, we do not reject the null hypothesis.
7. The median of differences is not significantly different from zero (5% level).

Assumption: The differences are continuous.

11.5   Sign Test Using Minitab for Windows

The procedure for performing the sign test using Minitab for Windows is shown in Table 11.2, which also includes the output.

In fact, Minitab uses the large-sample approximation test shown in Section 11.4, so the conclusions from this Section (11.5) might have been a little different from those in Sections 11.2 and 11.3, but they are in complete agreement as we shall see in what follows.

Example

Table 11.2   Minitab Input and Output for Sign Test for Population Median (Data from Section 11.2)

Input

Enter   200   550   290   170   180   350   190   210   160   250   into C1

Choose **Stat > Nonparametrics > 1-sample sign**
Enter **C1** in **Variable** box
Choose **Test Median**
Enter **250** in **Test Median** box
Choose **not equal to** in **Alternatives** box
Click on **OK**

Output

SIGN TEST OF MEDIAN = 250.0 VERSUS N.E. 250.0

	N	BELOW	EQUAL	ABOVE	P-VALUE	MEDIAN
C1	10	6	1	3	0.5078	205

Since $0.5078 > 0.05$ we do not reject the null hypothesis. This agrees exactly with the conclusion we reached in Section 11.3, where a probability value of 0.2539 was compared with 0.025. I say 'exactly' because 0.5078 is exactly twice 0.2539, and 0.05 is exactly twice 0.025.

11.6   Wilcoxon Signed Rank Test for the Median of a Population of Differences, 'Paired' Samples Data

In the **Wilcoxon signed rank test**, the null hypothesis tested is the same as for the sign test. Since the former uses the magnitudes as well as the signs of the differences, **it is more powerful than the sign test**, and hence is the preferred method when the magnitudes are known.

The general method for obtaining the calculated test statistic for the Wilcoxon signed rank test is as follows:

1. Disregarding **ties** (a tie means a difference of zero), the remaining n differences are **ranked** without regard to sign.
2. The sum of the ranks of the positive differences, T_+, and the sum of the ranks of the negative differences, T_-, are calculated. The smaller of the two is the calculated test statistic, T. A useful check is that $T_+ + T_- = \frac{n(n+1)}{2}$.

Example

Using the data of Section 11.3, the differences are as follows:

Differences (N score − S score)	7	−2	6	4	22	15	−5	1	12	15
Ranking the differences without regard to sign	1	−2	4	−5	6	7	12	15	15	22
The corresponding ranks are	1	2	3	4	5	6	7	$8\frac{1}{2}$	$8\frac{1}{2}$	10

Observe the example of tied ranks. The two values in rank positions 8 and 9 are equal (to 15), and are both given the mean of the ranks they would have had if they had differed slightly.

Now we calculate the value of the test statistic T.

T_+ = sum of the ranks of the + differences

$= 1 + 3 + 5 + 6 + 7 + 8\frac{1}{2} + 8\frac{1}{2} + 10 = 49$

T_- = sum of the ranks of the − differences

$= 2 + 4 = 6$

Since $n = 10$, $\frac{n(n+1)}{2} = \frac{10 \times 11}{2} = 55$.

Also, $T_+ + T_- = 49 + 6 = 55$, so this agrees.

The smaller of 49 and 6 is 6, so $T = 6$.

Setting out the seven-step method:

1. H_0: The median of the population of differences is zero, which implies that the median of N scores is equal to the median of the S scores.
2. H_1: The median of N scores is greater than the median of S scores.
3. 5% significance level.
4. *Calc T* = 6, from above.
5. *Tab T* = 10, from Table C.7 of Appendix C for a 5% significance level, one-sided alternative hypothesis, and $n = 10$.
6. Since *Calc T* < *Tab T* is true here, reject the null hypothesis.
7. The median of N scores is significantly greater than the median of S scores (5% level).

   Assumption: The distribution of the differences is continuous and symmetrical.

   (a) In step 6 we reject H_0 if *Calc T* ≤ *Tab T*, i.e., even if *Calc T* = *Tab T*.

   (b) When $n > 25$, Table C.7 cannot be used. Instead we use the method of Section 11.7.

   (c) The same data have been analysed using both the sign test and the Wilcoxon signed rank test. However, the conclusions are not the same! Using the sign test H_0 was not rejected (although the decision was a close one), while using the Wilcoxon test the null hypothesis was rejected. Since, as we have already mentioned, the latter test is more powerful, the latter conclusion is preferred.

11.7 Wilcoxon Signed Rank Test for Large Sample Sizes ($n > 25$)

When $n > 25$, Table C.6 cannot be used. Instead we use a normal approximation method by putting:

$$\mu_T = \frac{n(n+1)}{4} \quad \text{and} \quad \sigma_T = \frac{\sqrt{n(n+1)(2n+1)}}{24} \quad (11.2)$$

Example

Suppose that for $n = 30$ paired samples,

$$T_+ = 300, \quad T_- = 165, \quad \text{so that} \quad T = 165.$$

1. H_0: The median of the population of differences is zero.
2. H_1: The median of the population of differences is not zero (two-sided).
3. 5% significance level.
4. In order to obtain the value of the test statistic, we need values for μ_T and σ_T.

$$\mu_T = \frac{30 \times 31}{4} = 232.5, \quad \sigma_T = \sqrt{\frac{30 \times 31 \times 61}{24}} = 48.6$$

The normal distribution with these parameters is shown in Fig. 11.3. We can now obtain *Calc z*:

$$Calc\ z = \frac{T - \mu_T}{\sigma_T} = \frac{165.5 - 232.5}{48.6} = -1.38.$$

Note the use of the continuity correction as in Section 11.4.

5. *Tab z* = 1.96 from Table C.3(b), since this value of z corresponds to a tail area of 0.025, i.e., the significance level divided by 2 because the alternative hypothesis is two-sided.
6. Since $|Calc\ z| < Tab\ z$, we do not reject H_0.
7. The median of differences is not significantly different from zero (5% level).

Assumption: The distribution of differences is continuous and symmetrical.

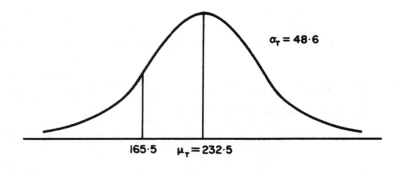

Figure 11.3   A Normal Distribution With $\mu_T = 232.5$, $\sigma_T = 48.6$

11.8   Wilcoxon Signed Rank Test Using Minitab for Windows

The procedure for performing a Wilcoxon signed rank test using Minitab is as in the following example, using the same data as in Section 11.6 (originally from Section 9.10). Once again, Minitab uses the large sample approximation test described in Section 11.7, even for small samples.

Example

Table 11.3   Minitab Input and Output for Wilcoxon Signed Rank Test for Median of a Population of Differences

Input

Enter the 10 differences: 7   −2   6   4   22   15   −5   1   12   15
into C1.
Choose **Stat > Nonparametrics > Wilcoxon**
Enter **C1** in **Variable** box
Choose **Test Median**
Enter **0** in **Test Median** box
Choose **greater than** in **Alternatives** box
Click on **OK**

Output

TEST OF MEDIAN = 0 VERSUS MEDIAN G.T. 0

	N	N FOR TEST	WILCOXON STATISTIC	P-VALUE	ESTIMATED MEDIAN
C2	10	10	49.0	0.016	7.0

The conclusion from this test is that H_0 should be rejected, since 0.016 is less than 0.05. This is the same decision as reached in Section 11.6, using the 'T' method.

11.9 Mann-Whitney U Test for the Difference Between the Medians of Two Populations, 'Unpaired' Samples Data

If we cannot justify the assumptions required in the unpaired samples t test (Section 10.14), the Mann-Whitney U test may be used for the following null and alternative hypotheses:

H_0: The two populations have distributions which are identical in all respects.

H_1: The two populations have distributions with different medians, but are otherwise identical.

The alternative hypothesis is two-sided here, but one-sided alternatives can also be specified.

The general method of obtaining the calculated test statistic for the Mann-Whitney U test is as follows. Letting n_1 and n_2 be the sizes of the samples drawn from the two populations, the $(n_1 + n_2)$ sample observations are **ranked** as one group.

Let the sum of the ranks of the observations in the sample of size n_1 be R_1, and let the sum of the ranks of the observations in the sample of size n_2 be R_2. Then U_1 and U_2 are calculated using Formulae (11.3a) and (11.3b).

$$U_1 = n_1 n_2 + \frac{1}{2} n_1 (n_1 + 1) - R_1 \qquad (11.3a)$$

$$U_2 = n_1 n_2 + \frac{1}{2} n_2 (n_2 + 1) - R_2 \qquad (11.3b)$$

(A useful check is $U_1 + U_2 = n_1 n_2$). The smaller of U_1 and U_2 is the calculated test statistic, U.

Example

As part of an investigation into factors underlying the capacity for exercise, a random sample of 11 factory workers took part in an exercise test. Their heart rates in beats per minute at a given level of oxygen consumption were as follows:

112 104 109 107 149 127 125 152 103 111 132

A random sample of 9 racing cyclists also took part in the same exercise test, and their heart rates were

91 111 115 123 83 112 115 84 120

These data are plotted in Fig. 11.4, which is similar to a dotplot but with the dots replaced by the actual heart rates to facilitate ranking. If we plotted the data on two dotplots, neither would look convincingly normal. A Mann-Whitney U test is appropriate here.

Ranking all 20 observations as one group, giving equal heart rates the average of the ranks they would have had if they had differed slightly, we obtain Fig. 11.5, e.g., 1 factory worker and 1 cyclist each had a heart rate of 111; these two values are in rank positions 8 and 9, so each is given the average rank of $(8 + 9)/2 = 8\frac{1}{2}$.

We now calculate U as follows:

$$n_1 = 11, \qquad n_2 = 9,$$

$$R_1 = 4 + 5 + 6 + 7 + 8\frac{1}{2} + 10\frac{1}{2} + 16 + 17 + 18 + 19 + 20 = 131.$$

$$R_2 = 1 + 2 + 3 + 8\frac{1}{2} + 10\frac{1}{2} + 12\frac{1}{2} + 12\frac{1}{2} + 14 + 15 = 79.$$

$$U_1 = 11 \times 9 + \frac{1}{2} \times 11 \times 12 - 131 = 34.$$

$$U_2 = 11 \times 9 + \frac{1}{2} \times 9 \times 10 - 79 = 65.$$

(Check: $U_1 + U_2 = 34 + 65 = 99$, $n_1 n_2 = 11 \times 9 = 99$, which checks.)

Factory workers, ($n_1 = 11$)	103,104,107,109,111,112		125,127 132 149,152
Cyclists ($n_2 = 9$)	83,84 91	111,112,115 120 123 115	

Figure 11.4 Heart Rates of Factory Workers and Cyclists

Factory workers ($n = 11$)		$4,5,6,7,8\frac{1}{2},10\frac{1}{2}$	16,17 18 19,20
Cyclists ($n = 9$)	1,2, 3	$8\frac{1}{2},10\frac{1}{2},12\frac{1}{2}$ 14,15 $12\frac{1}{2}$	

Figure 11.5 Ranks of Heart Rates of Factory Workers and Cylists

The smaller of U_1 and U_2 is 34, so $U = 34$ will be used in the seven-step method below:

1. H_0: The populations of the heart rates for factory workers and cyclists have identical distributions.
2. H_1: The distributions have different medians, but are otherwise identical (two-sided).
3. 5% significance level.
4. *Calc U* = 34, from above.
5. *Tab U* = 23, from Table C.8 of Appendix C for a 5% significance level, two-sided H_1, $n_1 = 11$, $n_2 = 9$.
6. Since *Calc U* > *Tab U* do not reject the null hypothesis.
7. The median heart rates for factory workers and cyclists are not significantly different (5% level).

Assumption: The variable is continuous. Since the number of beats per minute is large and may be the average of several observations, this assumption is reasonable in this case.

Notes

In Step 6 we reject the null hypothesis if *Calc U* ≤ *Tab U*, i.e., even if *Calc U* = *Tab U*.

When n_1 or n_2 is greater than 20, Table C.8 cannot be used. Instead, we use the method of Section 11.10.

11.10 Mann-Whitney U Test for Large Sample Sizes (n_1 or n_2 > 20)

When n_1 or n_2 > 20, we use the normal approximation method by putting:

$$\mu_U = \frac{n_1 n_2}{2} \quad \text{and} \quad \sigma_U = \sqrt{\frac{n_1 n_2 (n_1 + n_2 + 1)}{12}} \qquad (11.4)$$

Example

Suppose that for two unpaired samples of size $n_1 = 25$, $n_2 = 30$, we obtain $R_1 = 575$, $R_2 = 965$, $U_1 = 500$, $U_2 = 250$, so $U = 250$.

1. H_0: The two populations have identical distributions.
2. H_1: The two populations have distributions with different medians, but are otherwise identical (two-sided).

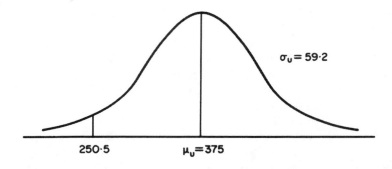

Figure 11.6   A Normal Distribution With $\mu_U = 375$, $\sigma_U = 59.2$

3. 5% significance level.
4. We calculate

$$\mu_U = \frac{25 \times 30}{2} = 375, \qquad \sigma_U = \sqrt{\frac{25 \times 30 \times 56}{12}} = 59.2.$$

The normal distribution with these parameters is shown in Fig. 11.6. We can now calculate the required test statistic:

$$Calc\ z = \frac{U - \mu_U}{\sigma_U} = \frac{250.5 - 375}{59.2} = -2.10.$$

Note the use of the continuity correction as in Section 11.4.
5. *Tab z* = 1.96, from Table C.3(b), this value of *z* corresponds to a tail area of 0.05/2, the significance level divided by 2, Since H_1 is two-sided.
6. Since $|Calc\ z| > Tab\ z$, reject the null hypothesis.
7. The medians are significantly different (5% level).

Assumption: The variable is continuous.

11.11   Mann-Whitney U Test Using Minitab

The procedure for performing a Mann-Whitney U test using Minitab is shown in the following example, using the same data as in Section 11.9. As expected, Minitab uses the large sample approximation test described in Section 11.10, even for small samples.

Example

Mann-Whitney U test for the data in Section 11.9 (heart rates for factory workers and cyclists):

Enter the 11 heart rates for the factory workers into C1, and the 9 heart rates for the cyclists into C2.

Choose **Stat** > **Nonparametrics** > **Mann-Whitney**
Enter **C1** in **first sample** box
Enter **C2** in **second sample** box
Choose **not equal to** in **Alternatives** box
Click on **OK**

Table 11.4   Minitab Output for Mann-Whitney U Test for the Difference Between the Medians of Two Populations

---

Mann-Whitney Confidence Interval and Test
C1   N = 11   MEDIAN = 112.0
C2   N = 9   MEDIAN = 112.0
Point estimate for ETA1 − ETA2 is 14.00
95.2 Percent C. I. for ETA1 − ETA2 is (−6.01, 32.01)
  W = 131.0
Test for ETA1 = ETA2 vs. ETA1 $\sim=$ ETA2 is significant at 0.2545.
The test is significant at 0.2543 (adjusted for ties)
Cannot reject at alpha = 0.05

A short explanation of Table 11.4 might be useful to some readers! ETA1 and ETA2 refer to the population medians. The first part of the output is concerned with a confidence interval for the difference between these medians (recalling Chapter 9). We are more interested in performing a hypothesis test here, but we could just remember that the fact that the C.I. contains zero implies that we would not reject H_0. The output also states W = 131.0, which was our value of R_1 in Section 11.9. Another small point is that $\sim=$ means 'not equal to' in Minitab notation. The two values 0.2545 and 0.2539 are p values; the second is more accurate than the first. So p value = 0.2539 and, since this is greater than 0.05, we cannot reject H_0 at the 5% level of significance, which is what the final line of the output is intended to convey.

11.12   Summary

Three nonparametric tests, namely, the sign test, the Wilcoxon signed rank test, and the Mann-Whitney U test are described for small and large sample cases. These tests require less rigorous assumptions than the corresponding *t* tests, but are less powerful if the assumptions of the *t* tests are valid.

Worksheet 11: Sign Test, Wilcoxon Signed Rank Test, Mann-Whitney *U* Test

Questions 1, 2, and 3 are multiple choice. Choose one of the three options in each case.

1. The Wilcoxon signed rank test is preferred to the *t* test when:
    (a) The sample sizes are large.
    (b) The data are paired.
    (c) The assumptions of the *t* test are invalid.
2. The sign test is
    (a) Less powerful than the Wilcoxon signed rank test.
    (b) More powerful than the paired samples *t* test.
    (c) More powerful than the Wilcoxon signed rank test.
3. The nonparametric equivalent of the unpaired samples *t* test is the:
    (a) Sign test.
    (b) Wilcoxon signed rank test.
    (c) Mann-Whitney *U* test.

Fill in the gaps in Questions 4, 5, 6, and 7.

4. Nonparametric tests are used to test_____ in cases where the _____ of the corresponding parametric tests are not valid.
5. However, when the _____ are valid, it is better to use parametric tests because they are more _____ than the corresponding nonparametric tests.
6. Power is the risk of rejecting the_____ hypothesis when the _____ hypothesis is correct. The _____ the power of a hypothesis test, the better.
7. The Mann-Whitney U test is a nonparametric test which corresponds to the _____ _____ *t* test. The latter is a more _____ test if two _____ are valid. These are that:
    (a) both variables are _____ distributed.
    (b) the _____ _____ are equal.
8. The sign test and the Wilcoxon signed rank test may both be used on paired samples data. Give examples of data which could:
    (a) Only be analysed using the sign test.
    (b) Be analysed using either test. Which test is preferable in this case?
9. Reanalyse the data from Worksheet 10, Question 12, using the sign test, and compare the conclusion with that of the *t* test.

10. What further information would you need, in addition to the data in Worksheet 10, Question 13, in order to carry out a sign test?

11. Reanalyse the data from Worksheet 10, Question 19, using the Wilcoxon signed rank test.

12. A psychologist tested 8 students, randomly chosen from the 11-year-old boys taught in the comprehensive schools of a city, using a standard aptitude test. The scores were

    135   103   129   96   122   140   110   91

    (a) Later the same subjects received a new (improved!) aptitude test, and the scores (in the same order of subject) were

        125   102   117   94   120   130   110   92

        Is there a significant difference between the average scores for the two tests? Use an appropriate nonparametric test.

    (b) Now assume that the scores in the second test refer to an independent second random sample of eight subjects. Is there a significant difference between the average scores for the two tests? Again use an appropriate nonparametric test.

13. An investigation was carried out on a trout farm to find the effect of a new feeding compound. Twenty fry (newly born fish) were randomly divided into two groups. Both groups were then kept under the same environmental conditions, but one group was fed with a standard feeding compound and the other group was fed with the new feeding compound. After a given period the fish were weighed. Their weights (in grams) were as follows:

Standard Compound	New Compound
510	521
507	476
490	489
496	512
523	521
508	498
534	505
497	547
491	542
506	492

Analyse these data using a nonparametric test.

14. Two brands of car tyres were tested in simulated road trials. The 'distances' travelled by 12 tyres of one brand and 12 tyres of the other brand before their treads had worn below the legal minimum limit were recorded to the nearest thousand kilometres:

Brand 1	47	44	39	41	39	42	51	44	55	45	49	46
Brand 2	43	33	40	38	31	39	34	40	35	37	38	32

Is one brand better than the other? Use a nonparametric test.

15. Look again at the 'Project' question (Worksheet 10, Question 25, which refers back to Worksheet 8, Question 10). Discuss whether any of the nonparametric tests covered in this chapter, namely Chapter 11, might be useful in assessing the effect of paid term-time employment on the academic performance of students.

Chapter 12

An Introduction to the Analysis of Variance (ANOVA)

12.1 Introduction

Chapter 10 described how to perform hypothesis tests, for example, how to compare the means of two populations when the sample data were unpaired (see Section 10.14), in which an unpaired samples t test to compare the A-level counts of 9 BA and 31 BSc students was discussed.

In this chapter we will analyse the same data in order to introduce a technique called Analysis of Variance, usually shortened to ANOVA. ANOVA can be used in a large number of situations which can broadly be described as 'the analysis of data from designed experiments and observational surveys'. In this chapter we will use ANOVA to compare two means so that we can see connections between the F test and the t test. In Chapter 15, we will use ANOVA in regression analysis, where part of the problem is to decide how much of the variation in one variable can be 'explained' by the variation in another variable.

12.2 An ANOVA Example

We met the term 'variance' in Section 4.12 where it was defined simply as the square of the standard deviation. Since standard deviation is a measure of variation, then variance is also such a measure, but the units in which it is measured will be strange, in the case of a continuous variable. For example, if our variable is height in cm, variance will have units cm-squared! Leaving that to one side, what formula should we use to calculate s-squared? One answer is found from Formula (4.3) by squaring both sides:

$$s^2 = \frac{\Sigma(x - \bar{x})^2}{n - 1}$$

In order to understand ANOVA, it is useful to think of variance as the ratio of what is known as the 'Sum of Squares' to the 'Degrees of Freedom'. We can see that $\Sigma(x - \bar{x})^2$ is the sum of squares of the n differences of x from the mean, $\bar{x}$, where n is the sample size, and we also know that $(n - 1)$ is the number of degrees of freedom when we obtain inferences for one sample of size n (see Section 9.7 and Section 10.10).

So how do we 'analyse the variance', when we perform an ANOVA? The word 'analyse' can mean 'break up into two or more separate parts'. In fact, in ANOVA, we analyse initially the Sum of Squares and then the Degrees of Freedom.

Consider the A-level counts of the BA and BSc students (see Table 12.1).

Table 12.1   A-level Counts of 9 BA and 31 BSc Students (Data from Table 1.1)

BA (x_1)	32	22	18	12	12	10	16	14	10						
BSc (x_2)	6	12	12	18	14	8	6	16	12	12	10	8	10	6	8
	4	6	12	8	22	8	6	18	10	10	12	10	10	8	6	24

Referring to the A-level counts of BA students as x_1 and the A-level counts of BSc students as x_2, we analyse the Total Sum of Squares into two parts in this particular example, which we call (a) the 'Between (type of degree) Sum of Squares' and (b) the 'Within (type of degree) Sum of Squares'. The formulae used to calculate these three measures are

$$\text{Total S.S.} = \Sigma x_1^2 + \Sigma x_2^2 - \frac{G^2}{(n_1 + n_2)}$$

where $G = \Sigma x_1 + \Sigma x_2$, the grand total   (12.1)

$$\text{Between S.S.} = \frac{(\Sigma x_1)^2}{n_1} + \frac{(\Sigma x_2)^2}{n_2} - \frac{G^2}{(n_1 + n_2)}$$ (12.2)

$$\text{Within S.S.} = \text{Total S.S.} - \text{Between S.S.}$$ (12.3)

Corresponding to the formulae for Sum of Squares, there are three more formulae for degrees of freedom:

Total d.f.	$= n_1 + n_2 - 1$	(12.4)
Between d.f.	$= 2 - 1 = 1$	(12.5)
Within d.f.	$=$ Total d.f. $-$ Between d.f.	(12.6)

We can think of $(n_1 + n_2 - 1)$ as 'the total number of observation -1', and we can think of '$2 - 1$' as 'the number of types of degree -1'.

Formulae (12.6) could also be written as:

Within d.f. $= (n_1 + n_2 - 1) - (2 - 1) = n_1 + n_2 - 2$
or simply Within d.f. $= (n_1 + n_2 - 2)$

In order to keep the algebra to a bare minimum, but at the same time giving some justification and understanding of the concept of ANOVA, we will apply the method (thus far) to a numerical example, namely, to the data in Table 12.1.

Example

Calculate the three Sums of Squares and the three Degrees of Freedom for the A-level count data of Table 12.1.

For BA students, $n_1 = 9$, $\Sigma x_1 = 146$, $\Sigma x_1^2 = 2772$.

For BSc students, $n_2 = 31$, $\Sigma x_2 = 332$, $\Sigma x_2^2 = 4240$.

So, $G = 146 + 332 = 478$,

$$\text{Between S.S.} = \frac{146^2}{9} + \frac{332^2}{31} - \frac{478^2}{40}$$
$$= 5924 - 5712$$
$$= 212.0 \ (4 \text{ s.f.})$$

$$\text{Total S.S.} = 2772 + 4240 - \frac{478^2}{40}$$
$$= 7012 - 5712$$
$$= 1300 \ (4 \text{ s.f.})$$

$$\text{Within S.S.} = 1300 - 212$$
$$= 1088 \ (4 \text{ s.f.})$$

$$\text{Between d.f.} = 2 - 1 = 1$$
$$\text{Total d.f.} = 9 + 31 - 1 = 39$$
$$\text{Within d.f.} = 39 - 1 = 38$$

We are now ready to draw up what is called the 'ANOVA table' for our example. All applications of ANOVA involve setting out such a table, the completion of which enables us to test null hypotheses of interest, using F tests.

Table 12.2 ANOVA Table for the Data in Table 12.1

Source of Variation	S.S.	d.f.	M.S.	Calc F
Between types of degree	212	1	212/1 = 212	212/28.63 = 7.40
Within types of degree	1088	38	1088/38 = 28.63	
Total	1300	39		

Notes

The 'Between S.S.' is a comparison of the A-level results of BA and BSc students. If the sample mean A-level count is the same for both types of degree, the Between S.S. would be zero. As an example, suppose that this mean is 10. Then Σx_1 would equal $9 \times 10 = 90$, $\Sigma x_2 = 31 \times 10 = 310$, $G = 90 + 310 = 400$, and so on. Between S.S. $= \frac{90^2}{9} + \frac{310^2}{31} - \frac{400^2}{40} = 900 + 3100 - 4000 = 0$.

The 'Within S.S.' is a measure of the variability within (inside) each of the two sets of data, measured separately about the mean for each set. If all the students taking the same type of degree have identical A-level counts, then the 'Within S.S.' would be zero.

As an example, suppose all BA students have a count of 15, while all BSc students have a count of 10. Then:

$$\Sigma x_1 = 9 \times 15 = 135, \quad x_2 = 31 \times 10 = 310, \quad G = 135 + 310 = 445.$$
$$\Sigma x_1^2 = 9 \times 15^2 = 2025, \quad x_2^2 = 31 \times 10^2 = 3100.$$

Hence Total S.S. $= 2025 + 3100 - \frac{445^2}{40} = 174.375$,

Between S.S. $= \frac{135^2}{9} + \frac{310^2}{31} - \frac{445^2}{40} = 174.375$,

and, finally, Within S.S. = Total S.S. − Between S.S. = 0, as expected.

Column 3 of the ANOVA table is headed M.S. which stands for **Mean Square**. Since the entries in this column are obtained by dividing the S.S. column by the d.f. column, it would have made more sense if Column 3 was headed **Variance**, looking back to the beginning of this Section!! I find it useful to think of Mean Square in ANOVA as a variance, and then it seems logical to use an F test on the ratio of the Between M.S. to the

Within M.S., since we introduced the F test in Section 10.15 to test the ratio of two variances. The appropriate formula in this ANOVA is

$$Calc\ F = \frac{\text{Between M.S.}}{\text{Within M.S.}} \quad \text{for 1} \quad \text{and} \quad n_1 + n_2 - 2 \quad \text{d.f.} \quad (12.7)$$

Here is the seven-step hypothesis test for the A-level problem:

1. $H_0: \mu_1 = \mu_2$, where μ_1 and μ_2 are the population means for the A-level counts of BA and BSc students, respectively.
2. $H_1: \mu_1 \neq \mu_2$, a two-sided alternative.
3. 5% significance level.
4. *Calc F* = 7.40, from Table 12.1 above.
5. *Tab F* = 4.10, from Table C.6 for 1, 38 d.f.
   The justification for the 1, 38 d.f. is that *Calc F* is obtained as the ratio of the Between M.S., which is associated with 1 d.f., to the Within M.S., which is associated with 38 d.f. To obtain *Tab F*, I need to locate 1 d.f. along the top of the F table for the relevant value of ν_1 (think of **top** of the ratio and **top** of the table). So, in this example, $\nu_1 = 1$. It follows that I must look for the d.f. for the 'denominator' or 'bottom' of the ratio down the left-hand side of the table.
6. Since *Calc F* > *Tab F*, $H_0: \mu_1 = \mu_2$ is rejected.
7. We conclude that there is a significant difference between the mean A-level counts of BA and BSc students (5% level of significance). Although the alternative hypothesis is two-sided, it is quite clear which type of degree student had the higher A-level count.
   From Section 9.11, the two sample means are 16.22 and 10.71 for BA and BSc students, respectively. We can now conclude that BA students have significantly higher A-level counts on average than BSc students (5% level of significance).

Assumptions: Both populations are normal and have the same variance. This was discussed in Sections 9.11 and 10.15 using the same data as in this example. We concluded then that both assumptions were justified. Moreover, statistical theory, the proof of which is beyond the scope of this book, indicates that the assumptions required for the F test in the ANOVA are not so critical as those required for the F test (as described in Section 10.15) for the equality of two variances.

A surprising aspect of the ANOVA we have just carried out, and I hope that you will find it surprising too(!), is that we have used a method called the analysis of **variance** to test a hypothesis about the difference between two **means**.

12.3 The Connection Between the Unpaired Samples *t* Test and the F Test in ANOVA

The same A-level count data were analysed in Section 10.14 using an unpaired samples *t* test, and now again in this chapter using ANOVA, which included an F test. You can now see that the first three steps of the seven-step method are the same for the two cases, as are the conclusions and assumptions stated after step 7. What about steps 4, 5, and 6?

Step 4. *Calc t* = 2.72, in the case of the *t* test.
Calc F = 7.40, in the case of the F test.
How are these connected? It isn't obvious is it?
The answer is that the square of *t* should equal F. For our example this means that 2.72^2 should equal 7.40, which is correct.

Step 5. *Tab t* = 2.02, for α = 0.025, and ν = 38.
Tab F = 4.10, for 5% significance, ν_1 = 1, and ν_2 = 38. As in step 4, the square of *t* equals F, since 2.02 × 2.02 = 4.1 approximately. The reason why we use α = 0.025 for the *t* test when H_1 is two-sided was explained in Section 10.6, but basically it is because the critical values of *t* can be positive or negative (see Fig. 10.2). For the F test, however, we note from Fig. C.4, for example, that F is always positive so we are only interested in the critical value of F which cuts off a right-hand tail area of 0.05 (assuming our significance level is 5%).

Step 6. For the *t* test, the 'decision rule' is:

$$\text{Reject } H_0 \text{ if } |Calc\ t| > Tab\ t$$

Applying this to the example, we rejected H_0 because 2.72 > 2.02. Similarly, for the F test, the rule was to reject H_0 if *Calc F* > *Tab F.* Applying this to the example, we rejected H_0 because 7.40 > 4.10. Since we know that the t^2 = F, it follows that in any given situation in which the data from two unpaired samples are analysed, the *t* test and the F test should always give exactly the same conclusion!

One of the assumptions of the *t* test was that the two populations have the same standard deviation and hence the same variance. For the A-level example, our estimate of this 'common variance' was 28.65 (see Section 10.14). It is no coincidence that, in the ANOVA table, the same number (or almost the same!) appears in the row labelled 'Within' and the column labelled M.S., recalling the point made earlier that a better name for 'Mean Square' is Variance.

The cynic might say: 'Why not forget ANOVA since we are more familiar with the t test, and the two tests reach the same conclusion in any case?

The answer is that although we would always prefer to use the t test to compare two means, there are many situations when the main source of variation, e.g., type of degree, has more than two levels or categories. In these cases, and others where there are more than two sources of variation, the t test is not appropriate. Instead, an F test, or several F tests, are called for. An example in which we wish to compare four means is set out in Section 12.4.

Finally, in this section, what do we do if the assumptions of the F test are not valid? The answer may be 'use a nonparametric test'. If the main source of variation has only two 'levels', the Mann-Whitney test is appropriate. When there are more than two levels, the Kruskal-Wallis test is used, but it is beyond the scope of this book (for details see *Statistical Methods in Psychology*, by D.C. Howell, 4th ed., Duxbury Press, London, 1997).

12.4   ANOVA to Compare Four Means, an Example

Example

Suppose that having high blood pressure means that a patient is more likely to suffer from heart disease in later life. Also suppose that we wish to compare four treatments A, B, C, and D, designed to lower the systolic aorta blood pressure. Suppose there are 18 patients, each being randomly assigned to one of the four treatments. Suppose further that the reduction in blood pressure (mm of Hg) for each patient was as follows:

Table 12.3   The Reduction in Systolic Aorta Blood Pressure for 18 Patients Receiving One of Four Treatments

Treatment			
A	B	C	D
35	30	25	16
35	28	22	15
28	28	19	15
27	25	18	
26	21		
24			

You might be tempted to do a number of t tests on these data, by considering all possible pairs of treatments. However, this approach is not appropriate for two reasons:

1. The six t tests you would need are not independent of one another, since each of the four sets of sample data are used in three of the six tests.
2. If each t test uses a 5% significance level, then the overall level for all six tests is considerably greater than 5%, although it is NOT $6 \times 5 = 30\%$.

The overall significance level can be reduced in importance by using a lower level of significance, 1% say, in each t test, but the lack of independence is still a problem, which is why we use the F test!

With such a small amount of data, it is difficult to check the two assumptions of ANOVA, which are as follows:

1. The fall in blood pressure 'within' each treatment is normally distributed.
2. The variances of the four distributions, one per treatment, are the same.

In theory, we could test each set of data for normality, using the Shapiro-Wilk test of Chapter 16. If that showed that the hypothesis of normality was not rejected, we could then check for 'equality of the four variances', using Bartlett's test (which is not covered in this book). With such small sample sizes, I do not believe these formal methods are any better than an 'eye-ball' test and/or a small amount of calculation.

For example, the summary statistics for each treatment are.

	A	B	C	D
Mean	29.2	26.4	21	15.7
S.D.	4.7	3.5	3.2	0.6
Sample size	6	5	4	3

The standard deviations are very similar except for Treatment D, where the three observations were virtually the same, probably coincidentally (see Table 12.2). In a situation like this I would perform an F test, and if the conclusions about the rejection of the null hypothesis were not very clear-cut, I would perform a Kruskal-Wallis test as a backup. Here are the calculations required for the 'four treatment' example:

$$\Sigma x_a = 175 \qquad \Sigma x_b = 132 \qquad \Sigma x_c = 84 \qquad \Sigma x_d = 46$$

So,

$$G = 175 + 132 + 84 + 46 = 437$$
$$\Sigma x_a^2 = 5215 \qquad \Sigma x_b^2 = 3534 \qquad \Sigma x_c^2 = 1794 \qquad \Sigma x_d^2 = 706$$

Sum of Squares of 18 observations $= 5215 + 3534 + 1794 + 706$
$$= 11,249.$$

Total S.S. $= 11,249 - \dfrac{437^2}{18} = 11,249 - 10,609.4 = 639.6$

$$\text{Between Treatment S.S.} = \frac{175^2}{6} + \frac{132^2}{5} + \frac{84^2}{4} + \frac{46^2}{3} - 10,609.4$$
$$= 11,058.3 - 10,609.4$$
$$= 448.9$$

Hence, Within Treatments S.S. $= 639.6 - 448.9 = 190.7$
Also, Total d.f. $= (6 + 5 + 4 + 3) - 1 = 17$
Between Treatments d.f. $= 4 - 1 = 3$, since we are comparing four treatments.
Within treatments d.f. $= 17 - 3 = 4$.

Table 12.4 ANOVA Table for the Data in Table 12.2

Source of Variation	S.S	d.f.	M.S.	Calc F
Between Treatments	448.9	3	149.63	10.99
Within Treatments	190.7	14	13.62	
Total	639.6	17		

Setting out the various steps in the seven-step method:

1. $H_0: \mu_a = \mu_b = \mu_c = \mu_d$

   where μ_a, for example, means the population mean reduction in systolic blood pressure for patients under Treatment A, and so on.
2. H_1: not all the four means are equal (see notes below).
3. 5% level of significance.

4. *Calc F* = 10.99, from ANOVA table.
5. *Tab F* = 4.26, for 3, 14 d.f, and 5% significance.
6. Since *Calc F* > *Tab F*, reject H_0.
7. We conclude that not all four means are equal.

Notes

When we compare only two means, it is easy to interpret the rejection of the null hypothesis; one mean is significantly larger than the other, and which is which depends on the direction of the difference in the sample means, in the case of a one-sided H_1. With a two-sided H_1, it is even easier. We simply say the means are significantly different. With four means, if H_0 is rejected, there are several possibilities. For example, one mean may be significantly greater than the other three, the latter being close together (i.e., not significantly different). In another case there may be two groups, each consisting of two means, and so on.

12.5   A Posterior Test if H_0 is Rejected in ANOVA

One way to refine the conclusion following the rejection of H_0 in ANOVA is to carry out what is called a **posterior test** (also known as a 'post hoc' test, or a 'multiple comparison test'). Some statisticians think that such a test is dubious because it is very similar to comparing all possible pairs of means, a procedure which we have already stated is not appropriate. Personally, I think that such a test is useful, because you can obtain a better feel for your data, by looking at it in a different way. However, I do not necessarily take the conclusions 'as gospel', i.e., uncritically, since this can lead to ambiguities.

Here is an example of a posterior test for the 'four treatment' experiment described in the previous section: (This test is sometimes referred to as the Studentised Range Statistic test, or the SNK test, after Messers Student, Neuman, and Keuls.)

We start by writing down the treatment means in rank order, from lowest to highest:

	D	*C*	*B*	*A*
Mean	15.3	21.0	26.4	29.2
Rank of mean	1	2	3	4
Sample size	$n_d = 3$	$n_c = 4$	$n_b = 5$	$n_a = 6$

Then, for each pair of means we calculate the difference in the means, and the standard error (s.e.) of the difference, where s.e. is given by the following formula:

$$\text{s.e.} = \sqrt{\frac{\text{Residual M.S.}}{n_{ij}}} \tag{12.8}$$

where n_{ij} is the harmonic mean of the two sample sizes n_i *and* n_j. For example, in the comparison between the means for treatments D ($n_4 = 3$) and A ($n_1 = 6$), the harmonic mean is

$$n_{36} = \frac{2}{\frac{1}{3} + \frac{1}{6}} = 4$$

Columns 1 to 5 in the Table 12.5 should be clear. Column 6 is headed by 'q', which is the critical value of the Studentised Range Statistic (see Table C.9) for a 5% level of significance. For example, the means of treatments D and A are ranked 1 and 4, respectively, so p = (4 − 1) + 1 = 4. Also the residual d.f. = 14 in the ANOVA table in Section 12.4, so q = 4.111 for the comparison of D and A. In the last column in Table 12.5, the entries are either Sig. if the Column 5 entry is greater than q, meaning that there is a significant difference between the current pair of means; or NS, if the Column 5 entry is less than q, meaning that the difference between the means is not significant. All six tests are at the 5% level.

Table 12.5   Posterior Test for the 'Four Treatment Example'

Comparison	Difference in Means	Harmonic Mean	s.e.	Difference/s.e.	q	Conc.
D vs. A	13.9	4.00	1.85	7.50	4.111	Sig.
D vs. B	11.4	3.75	1.91	6.0	3.702	Sig.
D vs. C	5.7	3.43	1.99	2.9	3.033	NS
C vs. A	8.2	4.80	1.68	4.9	3.702	Sig.
C vs. B	5.4	4.44	1.75	3.1	3.033	Sig.
B vs. A	2.8	5.45	1.58	1.8	3.033	NS

The conclusions from these tests are fairly clear:

1. The means of A and B are both significantly different from the means of C and D, but not from each other.
2. The means of C and D are not significantly different.

These conclusions may be represented graphically as follows:

	D	C	B	A
Mean	15.3	21.0	26.4	29.2

The names of the treatments, A, B, C, and D are positioned approximately on a scale according to their means, and pairs of treatments which are not significantly different are connected by a straight line.

Assuming that the best treatment is the one which gives the greatest drop in blood pressure, we can conclude that 'Treatments A and B are jointly the best, and are both better than C and D; but A and B are equally good, as are C and D'.

N.B. We should remember that we have performed six hypothesis tests, in each of which there was a 5% risk of rejecting H_0 when H_0 is, in reality, true. This means that the overall risk of wrongly rejecting H_0 is much higher than 5%, although it is impossible to say exactly what the risk is because the six tests are not independent of one another.

Another point here is that, in two of the six tests, the decision to conclude 'Sig or NS' (significant or not significant) was very marginal. This was the case when we compared D with C, and also when we compared C with B. Had either of these conclusions gone 'the other way' it would have been more difficult to separate the treatments.

Finally, in this example, it should be pointed out that sample sizes of 6, 5, 4, and 3 are less likely to lead to the detection of a real difference between the means than if larger sizes are used. Added to this is the point that the four sample sizes should be equal, excluding dropouts, unless we have prior knowledge that the variability in blood pressure for one or more treatments is much greater than for the other treatments.

12.6 ANOVA to Compare Means, Using Minitab for Windows

We will show the example of the previous section, this time using Minitab. For that example we need ONEWAY ANOVA. Minitab employs this terminology because there is only one factor, namely, Treatments, to consider apart from the variation 'between' patients, 'within' a particular treatment. The variance due to the latter is sometimes referred to as either the 'residual' variance or the 'error' variance. I find the idea of a residual, i.e., that variance which is left unexplained when all other factors have been taken into account, to be a useful concept. Minitab, however, uses 'error' rather than 'residual'.

The first step in carrying out a oneway ANOVA is to enter the blood pressure data into C1, say, and then indicate in C2 which treatment each patient received using a numerical code such as Treatment A = 1, B = 2, C = 3, and D = 4. (So, C1 will have 18 rows filled with reduction in blood pressure values, while C2 will consist of six 1s, five 2s, four 3s, and three 4s.)

Then,

> Choose **Stat** > **Anova** > **Oneway**
> Enter **C1** in **Response** box
> Enter **C2** in **Factor** box
> Click on **OK**

Note

Use the command Oneway if all the responses, e.g., fall in blood pressure, have been 'stacked' into one column, as in this example. If the responses have been put into several different columns, the data are said to be unstacked, and in this case the command Oneway [Unstacked] is appropriate.

Table 12.6   Minitab Output from ANOVA of Blood Pressure Data Analysis of Variance On C1

SOURCE	D.F.	S.S.	M.S.	F	p
Level	3	448.9	149.6	10.99	0.001
ERROR	14	190.7	13.6		
TOTAL	17	639.6			
LEVEL	N		MEAN		STDEV
1	6		29.167		4.708
2	5		26.400		3.507
3	4		21.000		3.162
4	3		15.333		0.557

POOLED STDEV = 3.691

Table 12.6 agrees with the ANOVA earlier in this section, except for changes in the layout and notation. The extra column for the p-value is useful because we do not then need to look up *Tab F* to decide whether to reject H_0. We simply say 'reject H_0 at the 5% level, and also at the 1% level, and very nearly at the 0.1% level'.

Finally, we note that the estimate of the common standard deviation of the four populations is given by Minitab as:

$$\text{POOLED STDEV} = 3.691$$

This value is the square-root of the error (or residual) M.S., i.e., the square root of 13.6 (see ANOVA table).

12.7   Summary

The Analysis of Variance (ANOVA) is a general method of data analysis, applicable to data from designed experiments and surveys. In Chapter 12 the concepts of ANOVA were introduced, and applied to testing the equality of two means using unpaired samples data. The resulting F test was compared with the t test from Section 10.14. (The same sample data was used for the two tests.) It was shown that the conclusions from the tests were identical.

One advantage of the F test is that it could be extended to cases in which more than two means were being compared. However, it was inappropriate to use the t test method if more than two means were being compared.

The problem of post hoc testing, if the null hypothesis of the ANOVA was rejected, was discussed briefly.

Finally, an example of how Minitab can be used to do some of the necessary calculations was described.

Worksheet 12:   ANOVA

1.   Use the sample data from Worksheet 10, Question 21, to test the null hypothesis that the mean strengths of the two cements, A and B, are equal, by means of an F test. You may assume that the assumptions of the test are valid, since these have already been covered in the Solutions to Worksheet 10. Compare your F test statistics and ANOVA conclusion with the t test and its conclusion. Check your answers using only a calculator with those using Minitab for Windows.

2.   Using only the summary statistics from Worksheet 10, Question 22, test the null hypothesis that corner shops are charging the same on average as supermarkets for the standard basket, by means of an F test. You may assume that the assumptions of the F test are valid (as in Question 1 above). Compare your F test results and conclusion with those of the unpaired samples t test.

Note: It is not possible to answer this question using Minitab. Do you know why?

3. Using the sample data from Worksheet 10, Question 23, test the null hypothesis that there is no difference between the mean amount of vanadium for the two areas, A and B.

    Compare the F test results and conclusion with those of the *t* test. Explain why the tabulated value of F and *t*-squared are not equal in this case.

4. A total of 30 mothers, each of whom had an 18-month-old baby, agreed to take part in a study of children's development. The mothers were randomly divided into three groups of 10, that did not meet each other.

    Group 1 was given a simple instruction session on how best to provide a healthy diet for their children.

    Group 2 was given a session of the same length explaining how small children could be taught to read.

    Group 3 was given several sessions on how small children could be taught to read and was also given suitable teaching materials.

    About one-third of the families moved out of the area and could not be followed up but, at the age of 8 the remainder of the children were tested for reading ability and comprehension. The test scores were as follows:

Group 1	Group 2	Group 3
101	88	124
88	104	128
97	98	157
84	118	134
114	126	122
102		109
82		
128		

It is intended to analyse these data using ANOVA. Are the two assumptions required for the appropriate hypothesis test likely to be valid in this case? Use only simple plots and calculations to answer this question.

If the answer is 'Yes', perform an ANOVA and draw a conclusion. Also discuss what you think a posterior test would conclude. Again, use only simple methods to answer this question.

Chapter 13

Association of Categorical Variables

13.1 Introduction

The inferential methods discussed in Chapters 9 to 12 involved data for one variable measured on a number of 'individuals', where the variable was numerical (either continuous, discrete, or ranked). We now turn to data for categorical (i.e., nonnumerical) variables. You may wish to re-read Section 1.2 before proceeding. Also, instead of one-variable (univariate) data, we will discuss two-variable (bivariate) data.

So, in this chapter we will be concerned with investigations in which **two categorical variables** are recorded for a number of 'individuals'. Such data may be set out neatly in **two-way contingency tables** (you may wish to reread Section 3.5 before proceeding). Initially, we will try to decide whether the two variables are independent or whether they are associated, by performing a hypothesis test, for example, the **Chi-squared (χ^2) test for independence**.

13.2 Contingency Tables

Remember that a categorical variable is one which is not numerical, but can take 'values' which are categories or classes.

Example

Suppose we want to find the reaction of children and adults to a new flavour of ice cream which a manufacturer would like to introduce. One variable could be 'Reaction to new ice cream', while the categories, at their simplest, could be 'liked the flavour' and 'disliked the flavour'. The other variable could be 'Type of subject' with categories 'adult' and 'child'. Since we now have two variables, each having two categories, we can set out the numbers of individuals in each of the four (2 × 2) cross-categories. Suppose the result is Table 13.1:

Table 13.1   A 2 × 2 Contingency Table for the Reaction of 140 Adults and 130 Children to a New Flavour of Ice Cream

| | Type of Subject | |
Reaction	Adult	Child
Liked flavour	90	100
Disliked flavour	50	30

What conclusions can be drawn from these data? If we didn't know any statistics, we might still make a sensible statement, for example, by calculating the percentages of adults and children who liked the flavour. For adults, this is 90/140 expressed as a percentage, i.e., 64%, while for children the percentage is 77% (100/130). So, it appears that a higher percentage of children like the new ice cream, compared with adults. But, we **do** know some statistics, and we realise that we have sample data, and so we should be thinking in terms of a hypothesis test or a confidence interval. Maybe you can't say exactly what test to carry out, but you should be able to think of a null hypothesis. TRY!

13.3 χ^2 Test of Independence, 2 × 2 Contingency Table Data

We will carry out the usual seven-step method of hypothesis testing:

1. H_0: The variables 'Type of subject' and 'Reaction to ice cream' are independent, i.e., there is no association between them.
2. H_1: The variables are not independent, they are associated (two-sided).
3. 5% significance level.

4. We denote the 'observed' frequencies in Table 13.1 by O, while the 'expected' frequencies are denoted by E. If all four expected frequencies are greater than or equal to 5 (see note (a) below), then the calculated test statistic for a 2×2 contingency table χ^2 test is

$$Calc\ \chi^2 = \sum \frac{(|O - E| - \frac{1}{2})^2}{E} \tag{13.1}$$

The upper-case sigma (Σ) means that we are going to 'sum', i.e., add together, the contributions from each of the four cells of the table. The four expected frequencies are obtained by applying the following formula to each cell of the table in turn:

$$E = \frac{row\ total \times column\ total}{grand\ total} \tag{13.2}$$

N.B. It is important to note that the E values are the frequencies we would expect if, for the purposes of calculation only, we assume that the null hypothesis is true. I hope you can see from Formula (13.1) that large differences between the O and the E values for a particular cell in the contingency table will lead to high values for Calc χ^2. So, intuitively, large values of this statistic tend to lead to the rejection of H_0.

Table 13.2   Expected Frequencies for the Data in Table 13.1

Reaction	Subject		Totals
	Adult	*Child*	*Totals*
Like flavour	90 (98.5)	100 (91.5)	190
Disliked flavour	50 (41.5)	30 (38.5)	80
Totals	140	130	270

Table 13.2 shows an expanded version of Table 13.1 to include the expected frequencies (E), in parentheses, and next to the observed frequencies (O). Also included are the row and column totals and the Grand Total. For example, for the row 2, column 1 cell, the row total is 80, the column total is 140, and the grand total is 270, so, corresponding to an O value of 50, we have an E value of:

$$\frac{80 \times 140}{270} = 41.5.$$

Also in Formula (13.1), we see $|O - E|$, which means we take the difference $O - E$ and ignore the sign. For example, for row 2 and column 2, $O - E = 30 - 38.5 = -8.5$, but $|-8.5| = +8.5$. For the data in Table 13.2, apply Formula (13.1) which incorporates $\frac{1}{2}$ subtracted from $|0 - E|$. This is **Yates's Continuity Correction**.

$$
\begin{aligned}
Calc\ \chi^2 &= \frac{(|90 - 98.5| - \frac{1}{2})^2}{98.5} + \frac{(|100 - 91.5| - \frac{1}{2})^2}{91.5} \\
&+ \frac{(|50 - 41.5| - \frac{1}{2})^2}{41.5} + \frac{(|30 - 38.5| - \frac{1}{2})^2}{38.5} \\
&= \frac{(8.5 - 0.5)^2}{98.5} + \frac{(8.5 - 0.5)^2}{91.5} + \frac{(8.5 - 0.5)^2}{41.5} \\
&+ \frac{(8.5 - 0.5)^2}{38.5} = 4.55
\end{aligned}
$$

5. *Tab* χ^2 is obtained from Table C.9, and we enter the tables for $\alpha = 0.05$, since the significance level is 5%, even though the alternative hypothesis is two-sided. The formula for the number of degrees of freedom, for a contingency table with r rows and c columns, is $(r - 1)(c - 1)$. So, for a 2×2 table, d.f. $= (2 - 1)(2 - 1) = 1$, and

$$Tab\ \chi^2 = 3.84, \quad \text{for} \quad 1\ \text{d.f.} \quad \text{and} \quad \alpha = 0.05.$$

6. Since *Calc* $\chi^2 >$ *Tab* χ^2, i.e., $4.55 > 3.84$, the null hypothesis is rejected.

7. We conclude that there is significant association between Type of subject and Reaction to a new ice cream. The 'direction' of the association is clear if we look at individual O and E values in one or more cells of Table 13.2. For example, fewer than expected adults liked the new ice-cream flavour ($90 < 98.5$), which implies the opposite for children ($100 > 91.5$).

Notes

The following notes relate to the 'ice cream' example and the analysis of contingency table data in general. Please read them carefully!

(a) All the expected frequencies must be at least 5, otherwise the formula for the calculated test statistic, (13.1), may not apply. If you have one or more expected frequencies below 5, you may:

(i) perform a different test, namely the Fisher exact test, if you have a 2 ← 2 table (see Section 13.7), or

(ii) combine rows (if r, the number of rows, is two or more) or columns (if c, the number of columns, is two or more), **but only if it makes sense to do so**; more on this in Section 13.4.

(b) The observations in a contingency table must be independent. The best way to ensure this is by taking a random sample of individuals. An example of dependent data might be if each child in the ice-cream example was related to one of the adults, thus forming pairs of subjects.

(c) The observations in the contingency table must be **frequencies**, not percentages, proportions or measurements.

(d) The null hypothesis of independence can be expressed in terms of 'no difference'. So, we can say that independence is equivalent to H_0: There is no difference in the population between the proportion of adults who like the new ice cream flavour and the proportion of children who like the flavour. In terms of conditional probabilities:

$$H_0: \text{P(liked flavour} \mid \text{adult)} = \text{P(liked flavour} \mid \text{child)}$$

(e) The fact that we may conclude that there is significant association between the variables does not necessarily imply cause and effect (we will make similar statements in connection with significant correlation coefficients in Chapter 14). So, we cannot conclude that being a child causes an individual to like the flavour of the ice cream, any more than we can say that liking the flavour causes an individual to become a child!

(f) Minitab for Windows does not automatically supply a p value for the χ^2 test of independence as described in this chapter. The steps required to obtain a p value are given on page T-262 of the *Student Edition of Minitab for Windows*, by John McKenzie et al., Addison Wesley, Longman, Reading, MA, 1995.

The formula for *Calc* χ^2 has been given for a 2 × 2 contingency table. For the general r × c table (with r rows and c columns), the following applies:

$$Calc\ \chi^2 = \sum \frac{(O - E)^2}{E} \quad \nu = (r - 1)(c - 1) \tag{13.3}$$

13.4 χ^2 Test of Independence, 3 × 3 Table

Because so many points were discussed in the previous example, we will now consider another example, this time starting off with a 3 × 3 table.

Example

In a large factory, 60 workers were randomly selected and asked to give their opinion on a new pension scheme that their employers were hoping to introduce. Of 10 workers with a high income, 8 were in favour of the new scheme, 1 was undecided and 1 was against the new scheme; of 25 workers on an average income, the numbers in favour, undecided, and against were 7, 3, and 15, respectively. Of 25 workers on a low income, the corresponding numbers were 2, 10, and 13, respectively. Are the opinions of the workers independent of their incomes? (Although income is quantitative it is treated as categorical for the purposes of this example.)

Table 13.3 can be formed from the above information. E values (calculated as in Section 13.3) are given in parentheses to one decimal place, which is sufficiently accurate.

Table 13.3   A 3 × 3 Contingency Table for Income and Opinion of 60 Workers

| | Opinion on a New Pension Scheme | | | |
Income	In Favour	Undecided	Against	Totals
High	8 (2.8)	1 (2.3)	1 (4.8)	10
Average	7 (7.1)	3 (5.8)	15 (12.1)	25
Low	2 (7.1)	10 (5.8)	13 (12.1)	25
Totals	17	14	29	60

Since there are three E values less than 5 in Table 13.3, we should not apply the χ^2 test to these data as they stand. We can, however, combine the top two rows, as in Table 13.4. It makes sense to combine rows here because the three income categories are ranked, and also because the top row is where the low E values are. However, it would not make sense to combine rows 1 and 3. Nor would it be sensible to combine columns rather than rows, because we would still be left with one E value of less than 5.

Table 13.4   A 2 × 3 Contingency Table for Income and Opinion of 60 Workers

| | Opinion on a New Pension Scheme | | | |
Income	In Favour	Undecided	Against	Totals
High or Average	15 (9.9)	4 (8.2)	16 (16.9)	35
Low	2 (7.1)	10 (5.8)	13 (12.1)	25
Totals	17	14	29	60

Now all the E values in Table 13.4 are greater than 5, so we can proceed with the seven-step method:

1. H_0: Opinion (on a new pension scheme) and income are independent
2. H_1: Opinion and income are associated (two-sided)
3. 5% significance level
4. The calculated test statistic for this 2×3 table is obtained by using Formula (13.3)

$$Calc \ \chi^2 = \frac{(15-9.9)^2}{9.9} + \frac{(4-8.2)^2}{8.2} + \frac{(16-16.9)^2}{16.9}$$
$$+ \frac{(2-7.1)^2}{7.1} + \frac{(10-5.8)^2}{5.8} + \frac{(13-12.1)^2}{12.1}$$
$$= 11.6.$$

5. *Tab* $\chi^2 = 5.99$ for $\alpha = 0.05$ and $\nu = (2-1)(3-1) = 2$.
6. Since *Calc* $\chi^2 > $ *Tab* χ^2 (11.6 > 5.99), we reject the null hypothesis.
7. We conclude therefore that opinion and income are significantly associated. Looking at the cells with the biggest contributions to the total of 11.6, we conclude that:
   (a) More of the high- or average-income workers are in favour than would be expected if income and opinion were independent.
   (b) More of the low-income workers are undecided than we would expect if income and opinion were independent.

Note

In this example the categories of the variable 'Opinion on a new pension scheme' are in a logical ranking order. In this situation, a further test, called a χ^2 **trend test**, is appropriate (see Section 13.7).

13.5 χ^2 Test of Independence, Using Minitab for Windows

To perform a χ^2 test of independence using Minitab for Windows, for example, on the data in Table 13.1, enter the frequencies 90 and 50 from column 1 of the contingency table into C1 of Minitab Data Window, and similarly with column 2 frequencies. Then:

Choose **Stat > Tables > Chisquare Test**
Enter **C1** and **C2** in the '**Columns containing the tables'** box
Click on **OK**

The output is shown below:

**Table 13.5   MTB > Chi-Square C₁ C₂. Expected counts
are printed below observed counts.**

	C_1	C_2	Total
Row 1	90	100	190
	98.2	91.48	
Row 2	50	30	80
	41.58	38.52	
Total	140	130	270

Note: Chi-sq. $= 0.737 + 0.793 + 1.749 + 1.8854 = 5.163$.
d.f. $= 1$.

The χ^2 value of 5.163 does not agree with the value of 4.55 obtained in Section 13.3. This is because Minitab uses Formula (13.3) instead of the more widely accepted Formula (13.1), which incorporates Yates's Continuity Correction (as in Section 13.3). My advice is to use Formula (13.3) for all except 2×2 tables, in which case use Formula (13.1). In the example from Table 13.2, 5.16 and 4.55 are both greater than 3.84, so the null hypothesis is rejected whether or not Yates' correction is used.

13.6 Fisher Exact Test

As stated in Note (a) after the example in Section 13.2, all expected frequencies must be at least 5, otherwise the χ^2 test may be invalid. If we are dealing with a 2×2 contingency table, we cannot combine rows or columns. However, there is another test we can perform instead, called the **Fisher exact test**. The null hypothesis is the same, but we do not calculate a test statistic as such. Instead, we calculate a probability, rather like we do in the sign test of Section 11.2, and compare with 0.05 if our test is at the usual 5% level. In more detail, the method is as follows:

Suppose that we have the following 2×2 table of observed frequencies a, b, c, and d, which give rise to four expected frequencies, at least one of which is less than 5:

a	b	a + b
c	d	c + d
a + c	b + d	n

Note that we have also included marginal row and column totals, and we let n stand for the sum of all the frequencies, so that $n = a + b + c + d$. We first calculate the probability

$$\frac{(a + b)!(c + d)!(a + c)!(b + d)!}{n!a!b!c!d!} \tag{13.4}$$

Assuming a two-sided alternative hypothesis, this procedure is repeated for all 2×2 tables with the same marginal totals. The sum of the probabilities so obtained will be 1 for obvious reasons. We require the sum of the initial probability (i.e., for the first table) and all other probabilities *which are less than or equal to the initial probabilty*. Calling this sum the 'total' probability, we reject the null hypothesis of independence if the total probability is less than 0.05, assuming a 5% level of significance.

Example

Forty students (see Table 1.1) were classified according to their sex and the type of degree for which they were studying (see Table 3.8 reproduced here as Table 13.6).

Table 13.6 Contingency Table for Sex and Type of Degree for 40 Students

	Type of Degree	
Sex	*BA*	*BSc*
Male	2	11
Female	7	20

We will test the null hypothesis that sex and type of degree are independent, against a two-sided alternative. Under H_0, the expected frequencies are as follows, using Formula (13.2),

$$\begin{array}{cc} 2.9 & 10.1 \\ 6.1 & 20.9 \end{array}$$

Since $2.9 < 5$, the χ^2 test is invalid, and a Fisher exact test is called for. For the data in Table 13.6, $a = 2$, $b = 11$, $c = 7$, $d = 20$, $n = 40$, and using Formula (13.4),

$$\text{Probability} = \frac{13!27!9!31!}{40!2!11!7!20!} = 0.253.$$

Since 0.253 is already greater than 0.05, the null hypothesis is not rejected, and there is no need to carry out any further probability calculations, which can only make the total probability larger. However, for illustration purposes only, the other nine tables with the same marginal totals and their corresponding probabilities (in parentheses) are

1	12	0	13	3	10	4	9	5	8	6	7	7	6	8	5	9	4
8	19	9	18	6	21	5	22	4	23	3	24	2	25	1	26	0	27
(0.105)		(0.017)		(0.310)		(0.211)		(0.083)		(0.018)		(0.002)		(0.000)		(0.000)	

(the total of these 10 probabilities is of course, 1, while the total of 0.253 plus all the others less than or equal to 0.253 is 0.690, which is, of course, still greater than 0.05).

The formal steps of the Fisher exact test for this example are

1. H_0: sex and type of degree are independent.
2. H_1: sex and type of degree are not independent (two-sided).
3. 5% significance level.
4. *Calc* probability = 0.69, or we could just say 'greater than 0.253', based on the initial table only.
5. 0.05 is the 'critical' probability (there is no 'tabulated' probability).
6. Since 0.69 or 0.253 > 0.05, the null hypothesis is not rejected.
7. We conclude that our data supports the hypothesis that sex and type of degree are independent. Perhaps a more useful conclusion is that the sex ratio (males/females) for BA students is the same as for BSc students.

Notes

As with the χ^2 test of independence, the observations must be independent. In addition, if H_0 is rejected, this does not imply cause and effect.

For a one-sided alternative hypothesis, we consider only those tables which are more extreme in the direction of the alternative hypothesis. For example, if the alternative hypothesis had been 'males are more likely to study for a BSc', we would consider only those tables with observed frequencies of 2, 1, and 0 in the top left cell of the table. This gives a total probability of 0.253 + 0.106 + 0.017 = 0.466 to be compared with 0.05 as above.

The Fisher exact test is not available on the student version of Minitab for Windows. This is a pity for two reasons. Although it can only be used for 2 × 2 tables, it is, as its name implies, an exact test. It can be used on all 2 × 2 tables, whether or not the lowest expected frequency is less than 5, and it will always give a more accurate answer than a χ^2 test of independence. The only reason why it isn't used routinely on 2 × 2 tables is because it is a more tedious calculation and because it is often missing from Statistical Computer Packages!

13.7 χ^2 Trend Test

This test was first mentioned in the note at the end of Section 13.4. It may be used in conjunction with the 'standard' χ^2 test for independence, described earlier in this chapter when one of the variables has more than two ordered categories. This was the case in Table 13.4, where the variable 'opinion on new pension scheme' had three ordered categories, namely, 'in favour', 'undecided', and 'against'. These categories are scored -1, 0, and $+1$, respectively, as shown in the following example. You should be able to follow the method even though formulae are not given, because that would make it look too 'mathematical' and possibly intimidating.

The rationale behind the trend test is that it takes account of the fact that the categories are ordered, which the 'standard' χ^2 test does not.

Example

Perform a χ^2 test on the data in Table 13.4.

The observed frequencies are

				Totals
	15	4	16	35
	2	10	13	25
Totals	17	14	29	60
writing down 'scores'	-1	0	$+1$	

we now calculate:

$$15 \times (-1) + 4 \times 0 + 16 \times (+1) = 1$$

$$17 \times (-1) + 14 \times 0 + 29 \times (+1) = 12$$

$$17 \times (-1)^2 + 14 \times (0)^2 + 29 \times (1)^2 = 46$$

$$Calc \ \chi_1^2 = \frac{60(60 \times 1 - 35 \times 12)^2}{35 \times 25(60 \times 46 - 12^2)} = \frac{7,776,000}{2,289,000} = 3.4$$

N.B. The '1' in the calculation of *Calc* χ_1^2 comes from the r.h.s. of the 'equals' sign three lines above.

We now calculate *Calc* $\chi_2^2 = 11.6$, using the standard Formula (13.3) for a 3 × 2 contingency table (Section 13.4). Finally, we calculate:

$$Calc\ \chi_3^2 = Calc\ \chi_2^2 - Calc\ \chi_1^2 = 11.6 - 3.4 = 8.2.$$

What do these three χ^2 statistics tell us? The first, $\chi_1^2 = 3.4$ tells us how much of the overall $\chi_2^2 = 11.6$ can be explained in terms of a linear trend as we go from the lowest category (of the variable with the ordered categories) to the highest category.

In our example, we compare 3.4 with 3.84 (from Table C.10 for $\alpha = 0.05$, and 1 d.f.). We conclude that there is no significant linear trend. Then we compare 8.2 (*Calc* χ_3^2) with 3.84, and we conclude that there is a significant nonlinear trend (checking on d.f., we have split the $(2 - 1)(3 - 1) = 2$ d.f. available for a 2 × 3 table into two cases of 1 d.f., enabling us to do two tests instead of one).

The absence of a linear trend can be supported by an eye-ball test looking at the proportions of those who are on a high or average income, who are in favour of the proposed new pension scheme. From Table 13.4, these proportions are 15/17 (88%), 4/14 (29%), and 16/29 (55%), so we can, albeit subjectively, see that there is no clear upward or downward linear trend in this case. Had the proportions for the second and third categories been in reverse order, i.e., so that the percentages were 88, 55, and 29, an eye-ball test would have been much more likely to support a linear trend. Of course, we would need to recalculate and test χ_1^2 and χ_3^2, noting that χ_2^2 would not change if the numbers in columns 2 and 3 of Table 13.4 were interchanged. This is left as an exercise for the reader.

The χ^2 trend test is not available on my version of Minitab for Windows.

13.8 Summary

In this chapter we looked at ways of testing whether two categorical variables were independent or associated. Sample data to test for independence, obtained from random samples of 'individuals', were set out as observed frequencies (O) in two-way 'contingency tables' with r rows and c columns. The null hypothesis of independence between the variables is tested by means of the χ^2 statistic using either Formula (13.1) for a 2 × 2 table, or Formula (13.4) for larger tables ($r > 2$ and/or $c > 2$).

The expected frequencies (E) are calculated using Formula (13.2) and are the frequencies we would expect, assuming independence. If any E value is less than 5, the Formulae (13.1) and (13.4) are invalid and alternative methods must be considered. These include combining rows or columns (if r > 2 or c > 2), or using the Fisher exact test in the case of 2 × 2 tables.

Rejection of the null hypothesis of independence does not necessarily imply cause and effect. If one of the variables has more than two categories, and they are logically ordered, a χ^2 trend test should be considered.

Worksheet 13: Association of Categorical Variables

Fill in the gaps in Questions 1 to 7.

1. A categorical variable can only take 'values' which are non-
....

2. If we collect data for two categorical variables for a number of 'individuals', the data may be displayed in a two-way or .... table. In such a table, the numbers in the various cells of the table are the number of .... in each cross-category and are referred to as .... frequencies.

3. The null hypothesis in the analysis of contingency table data is that: the two categorical variables are ....

4. In order to calculate the χ^2 statistic we first calculate the .... frequencies, using the formula:

$$E = \text{------------------}$$

5. If all the expected frequencies are greater than or equal to ......, the test statistic *Calc* χ^2 is calculated. Since the E values are calculated assuming the null hypothesis is true, high values of *Calc* χ^2 will tend to lead to the ....of the null hypothesis.

6. The number of degrees of freedom for *Tab* χ^2 are (......)(.......) for a contingency table with r rows and c columns, so for a 2 × 2 contingency table, the number of degrees of freedom is equal to ....

7. For a 2 × 2 contingency table, we reject the null hypothesis, at the 5% level of significance, if *Calc* χ^2 > ....

8. Choose one of the following three options, giving your reasons. The expected frequencies used in a χ^2 test on data from a contingency table must be (a) whole numbers, (b) all greater than or equal to 5, (c) greater than the corresponding observed frequencies.

9. Of 60 privately owned cars of a certain type and approximately the same mileage, 5 failed an M.O.T. of 40 similar 'company' cars, 9 failed the same test. Assuming that the 100 cars had been selected at random from all cars of the same type and mileage, test the

hypothesis that the proportion failing the M.O.T. is independent of whether they were privately owned cars or company cars.

10. For four garages in a city selling the same brand of unleaded four-star petrol the following table gives the number of male and female car drivers calling for petrol between 5 p.m. and 6 p.m. on a given day. Is there any evidence that the proportion of male to female varies from one garage to the other?

Sex of Driver	Garages				Totals
	A	B	C	D	
Male	25	50	20	25	120
Female	10	50	5	15	80
Totals	35	100	25	40	200

11. The examination results of 50 students, and their attendance (%) on a course, were as follows:

Attendance	Exam Result		Totals
	Pass	Fail	
Over 70%	20	5	25
30%–70%	10	5	15
Under 30%	5	5	10
Totals	35	15	50

Is good attendance associated with a greater chance of passing the examination?

12. Two types of sandstone were investigated for the presence of three types of mollusc. The numbers of occurrences were

Type of Sandstone	Type of Mollusc		
	A	B	C
Sandstone 1	15	30	12
Sandstone 2	15	0	6

Is there enough evidence to suggest that the proportions of the three types of mollusc is different for the two types of sandstone?

13. In a survey of pig farms it is suspected that the occurrence of a particular disease may be associated with the method of feeding.

Methods of feeding are grouped into two categories, A and B. Of five farms on which the disease occurred, four used method A and one method B. Of 15 farms on which the disease had not occurred, six used method B. Test for independence between the method of feeding and the occurrence of the disease.

14. Two drugs, denoted by A and B, were tested for their effectiveness in treating a certain common mild illness. Of 1000 patients suffering from the illness, 700 were chosen at random and given drug A, and the remaining 300 were given drug B. After 1 week, 100 of the patients were worse, 400 showed no change in their condition, and 500 were better. On the assumption that the two drugs are identical in their effect, complete a table similar in form to that below to show for each drug the expected number of patients getting worse, showing no change, and becoming better. The given table shows the observed number of patients in each category. Carry out a χ^2 test, at the 5% level, to determine whether the six observed frequencies are consistent with the assumption of identical effects. Also, carry out a χ^2 trend test, and state your overall conclusions.

	Number of Patients		
Drug Type	Becoming Worse	No Change	Becoming Better
Drug A	64	255	381
Drug B	36	145	119

15. The Admissions Tutor for a University Statistics course wanted to know whether interviewing applicants in terms 2 and 3 of an academic year would increase the proportion of applicants who, having been made conditional offers (dependent on A-level grades) by the university, actually enrolled on the course. Of a total of 68 applicants, all were asked to attend an interview at the university, but only 42 attended, of whom 9 actually enrolled on the course (those who did not attend for interview were given at least two additional alternative dates to choose from). Of the 26 (68 − 42) non-attenders, only one actually enrolled on the course.   Form a contingency table for these data, and test the hypothesis that interviewing and enrolment are independent. Assuming you reject the null hypothesis, what reservations do you have about concluding that interviewing increases the proportion of students enrolling?

Chapter 14

Correlation of Quantitative Variables

Besides, in many instances it is impossible to determine whether these are causes or effects

14.1 Introduction

In the previous chapter we discussed tests for the independence, or lack of association, of two categorical variables. If, instead, we are interested in **the association of two quantitative (numerical) variables** measured on a random sample of individuals from a population, we may:

(a) Summarize the sample data graphically in a scatter diagram (see Fig. 3.9), where the two variables are 'height' and 'distance from home';

(b) Calculate a numerical measure of the strength or degree of association, called a **correlation coefficient**;

(c) Carry out a test of the null hypothesis that there is no correlation in the bivariate population from which the sample data were drawn, and interpret the conclusion of this test with great care!

In case you are wondering why we seem to have switched from the word 'association' in Chapter 13 to the word 'correlation' in this chapter, the answer is that it is a convention to talk about association with respect to categorical variables. For quantitative variables, we use the word correlation conventionally and we can measure the strength of the correlation by means of a coefficient which we can calculate from sample data and then test its significance.

We will, in fact, discuss two such correlation coefficients, namely:

1. **Pearson's r**, which we will use if we can be reasonably sure that both of our variables are normally distributed,
2. **Spearman's r_s**, when we cannot assume normality for both variables, but we are able to rank each individual observation separately for each variable.

14.2 Pearson's Correlation Coefficient

Suppose we record the heights and weights of a random sample of six adult subjects (see Table 14.1). It is reasonable to assume that these variables are normally distributed, based on past experience of such variables, in which case Pearson's r is the appropriate measure of the strength of the association between height and weight. A scatter diagram of these data is shown in Fig. 14.1.

We will discuss the scatter diagram later, once we have calculated the value of Pearson's correlation coefficient, known as **Pearson's r** for short.

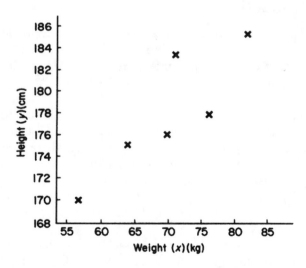

Figure 14.1 Scatter Diagram for the Heights and Weights of a Random Sample of Six Adults

Table 14.1 Heights (cm) and Weights (kg) of a Random Sample of Six Adults

Height	Weight
170	57
175	64
176	70
178	76
183	71
185	82

So lower-case r stands for the sample value of Pearson's coefficient, while we will use ρ (the Greek letter pronounced rho) for the population value. The formula for r is given below as Formula (14.1):

$$r = \frac{\Sigma xy - \dfrac{\Sigma x \Sigma y}{n}}{\sqrt{\left[\Sigma x^2 - \dfrac{(\Sigma x)^2}{n}\right]\left[\Sigma y^2 - \dfrac{(\Sigma y)^2}{n}\right]}} \tag{14.1}$$

where one of our variables is the x variable, the other is the y variable and n is the number of 'individuals' or 'subjects'. In correlation, it is an arbitrary decision as to which variable we call x and which we call y. Suppose we decide that weight is the x variable and height is the y variable as in Fig. 14.1. Then in Formula (14.1), Σx means the sum of the six weights, and so on. For the data in Table 14.1,

$$\Sigma x = 57 + 64 + 70 + 76 + 71 + 82 = 420$$
$$\Sigma x^2 = 57^2 + 64^2 + 70^2 + 76^2 + 71^2 + 82^2 = 29{,}786.$$
$$\Sigma y = 170 + 175 + \ldots = 1{,}067$$
$$\Sigma y^2 = 170^2 + 175^2 \ldots = 189{,}899$$
$$\Sigma xy = (57 \times 170) + (64 \times 175) + \ldots = 74{,}901$$

$n = 6$ individuals (or points on the scatter diagram).

$$r = \frac{74,901 - \frac{420 \times 1,067}{6}}{\sqrt{\left[29,786 - \frac{420^2}{6}\right]\left[189,899 - \frac{1,067^2}{6}\right]}}$$

$$= \frac{211}{\sqrt{386 \times 150.8}}$$

$$= 0.874.$$

How should we interpret a value for r, the sample correlation coefficient, of 0.874? In order to put this value into perspective, we can look at the scatter diagram of Fig. 14.1, where the general impression is of increasing weight to be associated with increasing height, and vice versa. Can you imagine a cigar shape round the six points, pointing neither horizontally nor vertically but at an angle (which will be highly dependent on our choice of scales for the two axes)?

More importantly, we can see that there is a trend from bottom left to top right, but it is not 'perfect' in the sense that, for a given weight, 70 kg, say, height can vary from, say, 174 to 184 cm. In fact, it can be shown that the value of Pearson's r would be exactly equal to 1 if we had the same trend, and all the points lay on a straight line. On the other hand, if there was a trend in which increasing one variable was associated with a *decrease* in the other variable, Pearson's r would be negative (between zero and -1) and would only take the lowest possible value i.e. if the points lay on a straight line.

If there is no tendency, and instead the points appear to be randomly distributed in the two-dimensional area of the scatter diagram, then r will be close to 0 (zero).

The three cases are shown in Fig. 14.2.

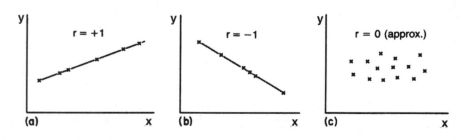

Figure 14.2 Scatter Diagrams for: (a) $r = +1$; (b) $r = -1$; (c) $r = 0$ (approx.)

Within the range of possible values for r from -1 to $+1$, we may describe a value of $+0.874$ (obtained above) as '**high positive correlation**'. But, a word of warning! Do not judge the association between two variables simply from the value of the correlation coefficient. We must also take into account the value of n, the number of 'individuals' contributing to the sample data. Intuitively, $r = 0.874$, based on a sample of 6 individuals, is not as impressive as $r = 0.874$ based on a sample of 60 individuals. Had we obtained the latter we would have much more evidence of the degree of association in the population. This intuitive argument is formalised in a hypothesis test for ρ, the population value of Pearson's correlation coefficient, in the next section.

14.3 Hypothesis Test for Pearson's Population Correlation Coefficient, ρ

Example

We will use the data and calculations of the previous section, and set out the seven-step method:

1. $H_0 : \rho = 0$. This implies that there is no correlation between the variables in the population.
2. $H_1 : \rho > 0$. This implies that there is a positive correlation in the population, i.e., increasing height is associated with increasing weight
3. 5% significance level.
4. The calculated test statistic is

$$Calc\ t = r\sqrt{\frac{n-2}{1-r^2}} \tag{14.2}$$

Notice that this formula contains n, the number of 'individuals' as well as r. For our data,

$$Calc\ t = 0.874\sqrt{\frac{6-2}{1-0.874^2}} = 3.61.$$

5. $Tab\ t = 2.132$ from Table C.5, for $\alpha = 0.05$, one-sided H_1, and $\nu = (n-2) = 6 - 2 = 4$, for this formula and these data, respectively. (It may help you to remember that the number of degrees of freedom, namely $(n-2)$ occurs in the formula for $Calc\ t$).

6. Since *Calc t* > *Tab t*, reject H_0.
7. There is significant positive correlation between height and weight.

Assumption: Height and weight are separately normally distributed.

Notes

There is a slightly shorter way of testing the null hypothesis if the population correlation coefficient is zero.
steps 1, 2, and 3 are the same as above; here are the other steps:

4. The calculated test statistic is simply *Calc r*, which for the example is equal to 0.874.
5. The tabulated test statistic is obtained from Table C.11, Critical values of Pearson's *r*. Since we have a one-sided alternative hypothesis, 5% significance level, and four degrees of freedom, we can read from Table C.11 that *Tab r* = 0.7293.
6. Since *Calc r* > *Tab r*, reject H_0
7. Step 7 is the same as for the *t* test above.

These two methods should always give exactly the same conclusion when applied to the same data. Which you choose is therefore up to you. *N.B.* You should read the next section before trying Worksheet 14.

14.4 The Interpretation of Significant and Nonsignificant Correlation Coefficients

The following six points, (a) to (f), should be considered whenever we try to interpret correlation coefficients:

(a) A significant value of *r* (i.e., when the null hypothesis, $H_0: \rho = 0$, is rejected) does not necessarily imply *cause and effect*. For the height/weight data, it clearly makes little sense to talk about 'height causing weight' or vice versa, but it might be reasonable to suggest that both variables are caused by (meaning 'depend on') a number of other variables such as sex, heredity, diet, exercise, and so on. For the kinds of example quoted regularly by the media, we must be equally vigilant. Claims such as 'eating animal fats causes heart disease', 'wearing a certain brand of perfume causes a person to be more sexually attractive', 'reducing inflation causes a reduction in unemployment' may or may not be true. They are virtually impossible to substantiate without controlling or allowing for many other factors which may influence the chances of getting heart disease, the level of sexual attraction, and the level of unemployment, respectively. Such careful research is difficult, expensive, and time-consuming, even in cases where the other factors may be controlled or allowed for. Where they may not be, it is misleading to draw confident conclusions.

(b) Pearson's correlation coefficient measures the *linear* association between the variables. So a scatter diagram may indicate nonlinear correlation, but the value of Pearson's *r* may be close to zero. For example, a random sample of 10 runners taking part in a local 'fun-run' of 10 miles may give rise to scatter diagram such as Fig. 14.3, if the time to complete the course is plotted against the age of the runner. A clear curvilinear relationship exists, but the value of Pearson's *r* would be close to zero.

(c) A few outlying points, called **outliers**, may have a disproportionate effect on the value of *r*, as in Fig. 14.4. In Fig. 14.4(a), the inclusion of the outlier would give a smaller value of *r* than if it were

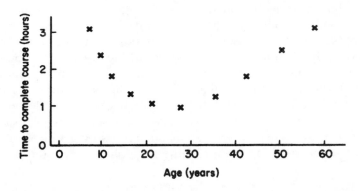

Figure 14.3   Scatter Diagram for Time to Complete Course and Age of Runner for a Sample of Ten

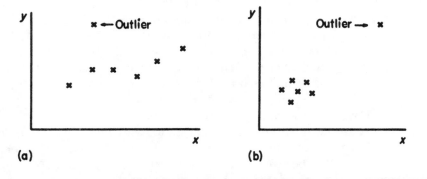

Figure 14.4   Two Scatter Diagrams, Each With an 'Outlier'

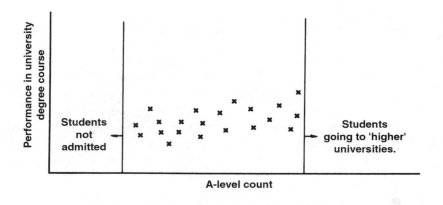

FIGURE 14.5   Scatter Diagram for A-level Count and University Performance

excluded from the calculations. In Fig. 14.4(b), the inclusion of the outlier would give a larger value of r.

In fact, in both cases the assumption that both variables are normally distributed looks suspect. In Fig. 14.4(a), the outlier has a value which is far away from the other values, and in Fig. 14.4(b), both the x and the y values of the outlier are extreme. However, we should not discard outliers simply because they do not fit into the pattern of the other points. We are justified in suspecting that some mistake may have been made in measuring and/or calculating the x and y values, or in plotting the point, or in some other way.

(d)   The value of r may be restricted, and may be nonsignificant in a hypothesis test, because the ranges of the x and y variables are restricted. For example, suppose the variables are 'a student's A-level count' and their subsequent 'performance in a degree course in a U.K. university'. The value of r for these variables in a particular university may be restricted by the fact that:

(i)   The university may require a minimum A-level count, such as 16 points (equivalent to BB or CCD).

(ii)   A student whose A-level count is well above that of the conditional offer made prior to A-levels may choose to go to a university higher up the pecking order. The value of r for students actually admitted to a particular university may be lower than if entry were unrestricted (see Fig. 14.5).

(e)   Nonsense correlations may result if two variables have increased or decreased in step over a period of time, but common sense indicates that the two variables are clearly unconnected. There are many examples of this type of correlation: the number of violent crimes and doctors' salaries may have increased over the last 10 years, and the correlation coefficient, for the 10 'individual years', may be significant. Clearly it would be nonsense to conclude that giving doctors more money results in more violent crime. Another nice example is the observation made in a Swedish town that in years when relatively more storks built their nests on house chimneys, relatively more babies were born in the town, and vice versa.

(f) Finally, in this section, if our sample size is too small we may not have enough data to detect a significant value of the correlation coefficient even if it exists in the population. On the other hand, if our sample size is too large, we may draw the conclusion that our correlation coefficient is 'significant' when its value is so small that it has no practical value. For example, when we have a sample of 102, a value of 0.2 for Pearson's r is significant at the 5% level, assuming a two-sided alternative hypothesis (see Table C.11 where $Calc\ r = 0.1946 < 0.2$). On the other hand, the same value of r is nowhere near significant for smaller sample sizes. (As we shall see in Chapter 15, a value of r of 0.2 means that one of our two variables 'explains only 4% of the variation in the other variable'.)

It may occur to you that, with all the reservations discussed above, there is little to be gained by calculating the value of a correlation coefficient and testing it for significance. The interpretation we can place on a significant value of r is that 'such a value is unlikely to have arisen by chance if there really is no correlation in the population, so it is reasonable to conclude that there is some correlation in the population'. In order to extend this conclusion to one of cause and effect, for example, requires much more information about other possible causal variables and consideration of the points made above in this section.

14.5   Spearman's Rank Correlation Coefficient

If two quantitative variables of interest are not normally distributed, Spearman's rank correlation coefficient may be calculated by **ranking** the sample data, separately for each variable, and using the formula:

$$r_s = 1 - \frac{6\Sigma d^2}{n^3 - n} \tag{14.3}$$

where r_s is the symbol for the sample value of Spearman's coefficient of rank correlation, and Σd^2 means the sum of the squares of the differences in the ranks of the n individuals. A nonparametric hypothesis test may then be carried out.

Table 14.2   Height (cm) and Weights (kg) Ranked for a Sample of Six Students

Height	Weight	Rank of Height	Rank of Weight	d^2
170	57	1	1	0
175	64	2	2	0
176	70	3	3	0
178	76	4	5	1
183	71	5	4	1
185	82	6	6	0
				$\Sigma d^2 = 2$

Formula (14.3) applies only when there are no '**tied ranks**'. A tie occurs when two or more of the sample values of a variable are equal and so are given the same rank. The calculation of Spearman's r_s in the case of tied ranks is discussed in Section 14.7.

Example (with no tied ranks)

For comparison purposes, the same data will be used as for the Pearson's r example of Section 14.2. The data are repeated in Table 14.2, which also shows the method for calculating Σd^2. We then calculate r_s as follows:

$$r_s = 1 - \frac{6 \times 2}{6^3 - 6}$$

$$= 0.943$$

What does a sample value of 0.943 for r_s tell us? Well, it can be shown that the possible range of values for r_s is -1 to $+1$ (the same as the range for Pearson's r). If $r_s = +1$, there is perfect agreement between the rankings of the two variables. If $r_s = -1$, there is perfect disagreement, (the highest rank for one variable corresponding to the lowest rank of the other variable, and so on). If $r_s = 0$, a particular rank for one variable may correspond to any rank of the other variable. So a value for r_s of 0.943 is high positive correlation (as we found for the same data when we calculated Pearson's r). Once again, though, this value should not be judged in isolation, since we must also take into account the number of 'individuals' n, which we do by carrying out a formal hypothesis test, which we now describe in Section 14.6.

14.6 Hypothesis Test for Spearman's Rank Correlation Coefficient

Example

Using the same data from the example in the previous section:

1. H_0:The ranks of height and weight are uncorrelated.
2. H_1:High ranks of height correspond to high ranks of weight (one-sided alternative).
3. 5% significance level.
4. *Calc* r_s = 0.943, from previous section.
5. *Tab* r_s = 0.829, from Table C.12 of Appendix C, for $n = 6$, one-sided alternative hypothesis and 5% level of significance.
6. Since *Calc* r_s > *Tab* r_s, reject H_0.
7. There is a significant positive correlation between the ranks of height and weight (5% level).

Assumption: We must be able to rank each variable.

The extensive notes in Section 14.4 on the interpretation of correlation coefficients apply equally to both the Pearson and the Spearman coefficients.

14.7 Spearman's Coefficient in the Case of Ties

In Section 14.4 it was stated that Formula (14.3) did not apply in the case of tied ranks. In this situation, we can either use a more complicated formula for r_s, or we can use the following ingenious method.

In the case of ties, calculate *Pearson's r* using the ranks rather than the original observed values of the two variables. It can be shown that the resulting value is the correct value of *Spearman's* r_s and this can then be tested for significance as in Section 14.6.

Example

A random sample of ten students were asked to rate, on a 10-point scale, two courses they had all taken. A rating of 1 means 'absolutely dreadful', while a rating of 10 means 'absolutely wonderful'. The data are given in the first two columns of Table 14.3. Here we are not interested in whether one course has a higher mean rating than the other (but, if we were, then a Wilcoxon signed rank test would be appropriate), but we **are** interested

Table 14.3 The Ranks of the Statistics and Mathematics Course Ratings of Ten Students

Statistics Rating	Mathematics Rating	Ranks of Statistics Rating (x)	Ranks of Mathematics Rating (y)
7	6	7	5.5
6	6	4.5	5.5
3	5	2	3.5
8	7	9	7.5
2	5	1	3.5
6	3	4.5	1
7	9	7	10
7	4	7	2
10	7	10	7.5
4	8	3	9

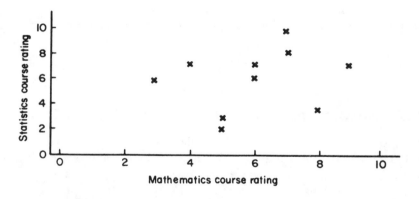

Figure 14.6 Scatter Diagram for the Ratings of Ten Students Taking Courses in Statistics and Mathematics

in whether there is a significant correlation between the ratings. In other words, do students who rate one course highly tend to rate the other course highly, relative to the ratings of other students, and vice versa? The scatter diagram, i.e., Fig. 14.6, indicates that the correlation coefficient may be positive but small, and we will hopefully confirm this subjective judgement when we calculate the sample value of Spearman's rank correlation coefficient.

For the ranks in Table 14.3, where x represents the ranks for the statistics course and y the ranks of the mathematics course, we have the following summary value

$$\Sigma x = 55 \qquad \Sigma y = 55$$
$$\Sigma x^2 = 382.5 \quad \Sigma y^2 = 383.5$$
$$\Sigma xy = 331.75 \qquad n = 10$$

Hence, Pearson's r is given by

$$r = \frac{331.75 - \frac{55 \times 55}{10}}{\sqrt{\left[382.5 - \frac{55^2}{10}\right]\left[383.5 - \frac{55^2}{10}\right]}}$$

$$= \frac{29.25}{\sqrt{80 \times 81}}$$

$$= 0.3634.$$

Since $Tab\ r_s = 0.648$ for $n = 10$ and a two-sided H_1, H_0 is not rejected, we conclude that the ranks of the two sets of ratings are not significantly correlated (5% level).

14.8 Correlation Coefficients Using Minitab for Windows

Minitab for Windows can give a scatter diagram using the following, assuming that we have our bivariate data into C1 and C2:

Choose **Graph** > **Character Graphs** > **Scatter Plot**
Enter **C1** in **Y variable** box
Enter **C2** in **X variable** box
Click on **OK**

A scatter plot (which is Minitab's name for a scatter diagram see Fig. 3.9) appears on the screen with the vertical axis labelled C1 and the horizontal axis labelled C2. Since this graph is using the same data as Fig. 14.1, it should be possible to conclude again that there is an approximate linear trend, which is what Pearson's r is measuring.

Also, we believe that height and weight are normal, so that Pearson's r is appropriate. In order to obtain Pearson's r:

Choose **Stat > Basic Statistics > Correlation**
Enter **C1 C2** in **Variables** Box
Click on **OK**
The Minitab output simply gives:
    Correlation of C1 and C2 = 0.874
This value agrees with the value quoted in Section 14.2 for these data.

Notes

1. Minitab does not perform a hypothesis test for ρ, the population value of Pearson's r.
2. Minitab does not calculate Spearman's r_s, except by the use of the ingenious method shown above, i.e., initially ranking each variable and then finding Pearson's r for the ranks.

Following on from Note 2 above, we will obtain Spearman's r_s for the height/weight data of Table 14.2 by getting Minitab to rank each variable and then use the Pearson formula on the ranks. The steps are as follows, assuming that the data are still in C1 and C2:
    Choose **Calc > Mathematical Expressions**
    Enter **C3** in **New Variable** box
    Enter **rank (C1)** in **Expression** box
    Choose **Calc > Mathematical Expressions**
    Enter **C4** in **New Variable** box
    Enter **rank (C2)** in **Expression** box
    Choose **Stat > Basic Statistics > Correlation**
    Enter **C3 C4** in **Variable** box
    The output should state:
    Correlation of C3 and C4 = 0.943,
as expected from the example in Section 14.5.

14.9 Summary

As in Chapter 13, inferences from bivariate sample data are discussed, but in this chapter the case in which the two variables are quantitative (rather than categorical) is covered.

The scatter diagram is a useful and important summary of this type of data. A measure of the degree of association between the variables is provided by a correlation coefficient. If both variables are normally distributed, Pearson's r is the appropriate coefficient. In other cases we may use Spearman's rank correlation coefficient, assuming the data are capable of being ranked.

Hypothesis tests may be used to test the significance of both coefficients, Pearson's test being the more powerful if both variables are 'normal'.

There are several important points to bear in mind when we try to interpret correlation coefficients.

Worksheet 14: Correlation of Quantitative Variables

Fill in the gaps in Questions 1 to 6.

1. If two quantitative variables are measured for a number of individuals the data may be plotted in a _____ _____

2. A _____ _____ is a measure of the degree of association between two quantitative variables.

3. If it is reasonable to assume that each of two variables is normally distributed and we wish to obtain a measure of the degree of linear association between them, the appropriate _____ _____ to calculate is _____'s and has the symbol ___. For the population the symbol is ___.

4. In calculating _____ we must decide which of our variables is the x variable and which is the y variable. However, the choice is

   _____.

5. The value of r (or r_s) must lie somewhere in the range ___ to ___. If the points on the scatter diagram indicate that, as one variable increases the other variable tends to decrease, the value of r will be _____. If the points show no tendency to either increase or decrease together, the value of r will be close to _____.

6. In order to decide whether there is a significant correlation between the two variables, we carry out a hypothesis test for the population parameter ____, if the variables can be assumed to be _____ _____. If we cannot make this assumption the null hypothesis is that the ranks of the two variables are _____.

Questions 7, 8, and 9 are multiple choice. Choose one of the three options in each case.

7. A correlation coefficient of 0.8 between two variables implies:
   (a) That as one variable increases the other decreases.
   (b) That $H_0 : \rho = 0$ should be rejected in favour of $H_1 : \rho > 0$.
   (c) Nothing, since there is insufficient information.

8. A random sample of 12 pairs of values have a Spearman rank correlation coefficient of 0.54. We can conclude that, for a 5% level of significance:
   (a) H_0 should be rejected in favour of a one-sided H_1.
   (b) H_0 should be rejected in favour of a two-sided H_1.
   (c) H_0 should not be rejected in favour of a one-sided H_1.

9. A significantly high negative value of a correlation coefficient between two variables implies:
   (a) A definite causal relationship
   (b) A possible causal relationship
   (c) That as one variable increases the other increases.

10. The percentage increase in unemployment and the percentage increase in manufacturing output were recorded for a random sample of ten industrialized countries over a period of a year. The data are listed below. Draw a scatter diagram. Is there a significant correlation? What further conclusions can drawn, if any?

Percentage Increase in Unemployment	Percentage Increase in Manufacturing Output
10	−5
5	−10
20	−12
15	−8
12	−4
2	−5
−5	−2
14	−15
1	6
−4	5

11. A company owns eight large hotels, one in each of eight geographical areas. Each area is served by a different commercial television channel. To estimate the effect of television advertising, the company carried out a month's trial in which the number of times a commercial advertising the local luxury hotel was shown was varied from one area to another. The percentage increase in the receipts of each hotel over the three months following the month's trial was also calculated:

Area	1	2	3	4	5	6	7	8
Number of times the commercial shown	0	0	0	10	20	30	40	50
Percentage increase in receipts	−2	5	10	5	7	14	13	11

What conclusions can be drawn?

12. In a mountainous region, a drainage system consists of a number of basins with rivers flowing through them. For a random sample of seven basins, the area of each basin and the total length of the rivers flowing through each basin are as follows:

Basin Number	Area (Sq. km)	River Length (km)
1	7	10
2	8	8
3	9	14
4	16	20
5	12	11
6	14	16
7	20	10

Are larger areas associated with longer river lengths?

13. From the data in the table that follows, showing the percentage of the population of a country using filtered water and the death rate due to typhoid for various years, calculate the correlation coefficient and test its significance at the 5% level. What conclusions would you draw about the cause of the reduction in the typhoid death rate from 1900 to 1912?

Year	Percentage Using Filtered Water	Typhoid Death Rate Per 100,000 Living
1900	9	36
1902	12	37
1904	16	35
1906	21	32
1908	23	27
1910	35	22
1912	45	14

14. A random sample of 20 families had the following annual income and annual savings in thousands of pounds (£):

Income	Savings
10.2	0.4
40.6	1.0
50.4	0.6
30.0	11.4
20.6	1.4
31.2	2.6
31.0	0.8
14.6	8.4
17.2	4.0
24.6	1.2
28.0	0.6
17.8	0.2
24.8	0.0
28.0	1.4
32.0	0.4
28.0	0.6
30.6	2.0
24.8	1.2
20.6	1.0
22.6	1.4

Is there a significant positive correlation between income and savings?

15. For the data in Table 1.1, which are the only two continuous variables? Which correlation coefficient should be calculated in order to measure the degree of association between them? Obtain this coefficient using the data from the first 10 students only, and test its significance at the 5% level.

16. (a) Think of some sample data, which might have been obtained from a bivariate population, calling the variables x and y, and assuming the sample size is $n = 5$, is such that Pearson's r is equal to (i) $+1$, (ii) -1, (iii) exactly 0.

    Check the values of r using Minitab.

    (b) Suppose that when you draw a random sample of size 5 from a bivariate population, where the variables are x and y, you find that the five points lie on a straight line parallel to the (i) x axis, (ii) y axis.

    Think of some data for each of these cases, then calculate Pearson's r for your data, using both a calculator and Minitab. Explain any inconsistencies in the answers you obtain.

17. Just as the previous question allowed you to have some individual input by providing your own data, so this question is to think

about a hypothetical project to see if you know how to 'design' such a project. Your design may be different from another student's, but may be equally valid. Read on....

Suppose you are an undergraduate student in a U.K. university and you are interested in, for example, the correlation between A-level scores and degree performance. If this particular example is not relevant to you, choose one that is by suggesting any two measures of academic achievement, let's call them x and y, measured at two points in time in your educational career. Now aim to get at least 20 pairs of values of x and y from 20 individuals (fellow students?) Before you collect any data, you will need to address a number of practical considerations. The following list of such considerations is not exhaustive; it is just to get you started!

(a) How will you ensure that your sample of individuals is randomly drawn from a population, and what is the population?

(b) How will A-level count be defined for students taking four A-levels? For example, will General Studies be included in your count? What about retakes? Are some A levels easier than others even in the same subject, for example modular A levels?

(c) What if a student took A levels 20 years ago; or has some Open University credits; or did some of the French Baccalaureate, or has studied beyond GCSE in Scotland or Ireland?

(d) What if a student has some post GCSE education, but does not have enough A-level points, but has significant life/job experience?

(e) How do you assign a number to 'degree performance', if a student's final transcript on graduation simply indicates one of the following five categories:
First-class honours,
Upper Second-class honours,
Lower Second-class honours,
Third-class honours,
Non-honours ('Ordinary degree')?

(f) What about the students who leave the course before graduating, for whatever reason?

Write a report on what you have learned so far in your research, before you collect any data.

Chapter 15

Regression Analysis, An Introduction

15.1   Introduction

When two quantitative variables are measured for a number of individuals we may be more interested in predicting the value of one variable from the value of the other variable, than in obtaining a measure of the degree to which the variables are associated. We discussed the latter in Chapter 14, and we will discuss the former in this chapter.

For example, if trainee salespeople take a test at the end of their training period, can these test scores be used to predict their first-year sales, and how accurate are these predictions? One way to answer such a question is to collect both the test scores and the first-year sales of a number of salespeople (who have already completed their first year). The next step is to draw a scatter diagram of these data. If the diagram indicates a possible linear trend, then, instead of calculating Pearson's r, we 'fit' a straight line though the 'data points'. If the equation of this line is obtained using Formulae (15.1) and (15.2) this line is called the **regression line**, and the equation of the line is called the **linear regression equation**.

As an aside, the word 'regression' comes from work done during the 19th century by Sir Francis Galton. He collected the heights of fathers and their sons and put forward the idea that since very tall fathers tended to have slightly shorter sons, and very short fathers tended to have slightly taller sons, over a number of generations there would be what Galton called a 'regression to the mean'.

15.2 Determining the Regression Equation, an Example

The linear regression equation is of the form:

$$y = a + bx \tag{15.1}$$

where x and y are our two variables, and a and b are, respectively, the **intercept** and the **slope** (or gradient) of the line. Let's look at the example of the salespeople with their test scores and first-year sales. The first thing to do is to decide which is the 'x variable' and which is the 'y variable'. You may remember that this was an arbitrary decision if we were interested only in correlation. This is not the case in regression analysis. In general, the y variable is the one we wish to predict, while x is the variable we use to predict y.

If the test taken by the trainees is a **good** predictor of first-year sales, then it can be used as part of the selection process of would-be salespeople. The aim of the test is to forecast (i.e., predict) the future sales. So it is more logical to call first-year sales 'the y variable', and test score is the 'x variable'.

But what does 'good' mean in the previous paragraph? One answer could be that the correlation between first-year sales and test score is high. But we are in a chicken-and-egg situation again (recalling Sections 9.6 and 9.9). We need a pilot survey in which we measure both the test scores of a sample of trainees at the end of their training period and we need to wait a year to obtain the first-year sales of the same people. Then we can obtain a model of these data, the simplest type being the linear model, which is simply an equation relating to x and y.

The general form of the so-called linear regression equation of 'y on x' is

$$y = a + bx$$

where a and b depend on our sample data. For the test score/first-year sales example we could write the regression equation as follows:

$$(\text{first-year sales}) = a + b \,(\text{test score})$$

Now we need sample data in order to obtain estimates a and b.

Suppose the data in Table 15.1 have been obtained from a random sample of eight salespeople:

Table 15.1   First-Year Sales (£ Thousands) and Test Scores of Eight Salespeople

First-Year Sales (y)	Test Score (x)
105	45
120	75
160	85
155	65
70	50
150	70
185	80
130	55

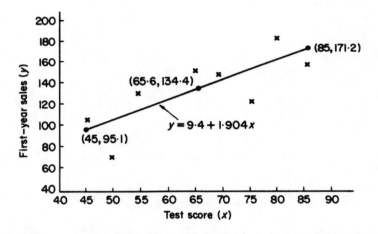

Figure 15.1   Scatter Diagram for First-Year Sales and Test Score

A scatter diagram of these data is shown in Fig. 15.1.
The impression given by the scatter diagram is of a fairly weak positive correlation. We obtain estimates a and b using the data in Table 15.1 and Formula (15.2), where n is the number of individuals (or the number of points on the scatter diagram).

$$b = \frac{\Sigma xy - \frac{\Sigma x \Sigma y}{n}}{\Sigma x^2 - \frac{(\Sigma x)^2}{n}} \quad \text{and} \quad a = \bar{y} - b\bar{x} \tag{15.2}$$

Where

$$\bar{x} = \frac{\Sigma x}{n} \quad \text{and} \quad \bar{y} = \frac{\Sigma y}{n} \tag{2.1}$$

For the Sales and Test scores example, we require the following summations (refer to Section 14.2 for similar calculations, if you need to).

$$\Sigma x = 525 \qquad \Sigma x^2 = 35,925$$
$$\Sigma y = 1,075 \qquad \Sigma y^2 = 153,575$$
$$\Sigma xy = 73,350 \qquad n = 8$$

Using these values, we obtain

$$b = \frac{73,350 - \frac{525 \times 1,075}{8}}{35,925 - \frac{525^2}{8}} = \frac{2,803}{1,472} = 1.904$$

$$a = \frac{1,075}{8} - 1.904 \times \frac{525}{8} = 134.4 - 125.0 = 9.4.$$

Putting these values into Formula (15.2) gives the linear regression equation for our data:

$$y = 9.4 + 1.904x, \text{ or}$$

$$(\text{first year sales}) = 9.4 + 1.904(\text{test score}).$$

It can be shown, using calculus, that this line has a special property which makes it in some sense the best straight line. Here 'best' implies that this line *minimises the sum of squares of the distances from the data points to the line in the y direction,* i.e., parallel to the y axis.

For this reason the line is sometimes referred to as the 'least squares regression line of y on x'.

15.3 Plotting the Regression Line on the Scatter Diagram

We can now plot the regression line on the scatter diagram. In theory, we could substitute any value of x into the regression equation and calculate the predicted value of y. In practice, however, we should not **extrapolate** our regression line. In other words, we should not use values of x outside the range of our sample data, because the line may not be valid in that case. For this reason, it is a good idea to choose the minimum and maximum values of x from our sample data and find the corresponding predicted values of y.

In our example, the minimum and maximum values of x are 45 and 85, respectively:

When $x = 45$, the predicted value of y is $9.4 + 1.904 \times 45 = 95.1$.

When $x = 85$, the predicted value of y is $9.4 + 1.904 \times 85 = 171.2$. Now we can plot the points (45, 95.1) and (85, 171.2) on the scatter diagram, as shown in Fig. 15.1., and join by a straight line.

As a check on the position of the regression line, it should pass through the point which is the 'centre' of the data, namely, $(\bar{x}, \bar{y})$. For the example, this point is (525/8, 1075/8) or (65.6, 134.4).

15.4 Predicting Values of y

We stated earlier that the main purpose of regression analysis was to predict y values from x values. We have already seen how to use the regression equation to do this for $x = 45$ and $x = 85$. Let's do another example, say, when $x = 60$.

Example

Predict first-year sales when the test score x equals 60.

The predicted y when $x = 60$ is equal to $9.4 + 1.904 \times 60 = 123.6$ (4 s.f.)

What does 'predicted $y = 123.6$ mean'? It means that 123.6 is our single-value estimate of the mean value of sales for all salespeople who achieved a score of exactly 60 in the test.

A single-value estimate is, as you might expect, a single number. Using the ideas of Chapter 9, we will also consider in this chapter obtaining an interval estimate, i.e., a confidence interval for predicted y, not just for one value of x, but for any value of x (within the range of x in the sample data).

We will also use hypothesis testing, the other main branch of statistical inference, to test the null hypothesis that the slope of the underlying population regression line is zero.

A good way to understand the ideas behind inferential methods in regression analysis is to use Analysis of Variance (ANOVA), which we first met in Chapter 12. Before we do that, however, we need to introduce the term **residuals** since these are at the centre of all inferences in regression.

Table 15.2   Predicted Values and Residuals for the Data Points in Fig. 15.1

First-Year Sales y	Test Score x	Predicted y	Residual
105	45	95.1	9.9
120	75	152.2	−32.2
160	85	171.2	−11.2
155	65	133.2	21.8
70	50	104.6	−34.6
150	70	142.7	7.3
185	80	161.7	23.3
130	55	114.1	15.9

15.5   Residuals

For each data point on the scatter diagram, the residual is given by the simple equation:

$$\textbf{residual} = \textbf{observed } y - \textbf{predicted } y$$

For example, the first data point in Table 15.1 is given by $x = 45$ and $y = 105$. The predicted value of y when $x = 45$ is 95.1, so the residual for this data point is $105 - 95.1 = 9.9$. This is the vertical distance from the data point to the regression line, and it is a positive number because the point is above the line. Data points below the regression line give rise to negative residuals. Table 15.2 gives predicted values and residuals for all eight data points in the sales/test score example.

Statistical theory indicates that the sum of the residuals should be zero, apart from rounding errors. In the above table, the sum of the residuals is 0.2, which is close to zero. There is interest, too, in the sum of the squares of the residuals, because you will remember that our regression line is such that it minimises this sum of squares. For the example, this sum of squares is equal to $9.9^2 + (-32.2)^2 + \ldots + 15.9^2 = 3782$. When we discuss ANOVA in the next section, we will return to the concept of the 'residual sum of squares' and the value of 3782 found in our example.

15.6   ANOVA in Regression Analysis

As we saw in the ANOVA of Chapter 12, we need to share out a 'Total Sum of Squares' between two or more 'Sources of Variation'. In the regression example we have been considering in this chapter, the total sum of squares refers to the variable y only. There are two sources to 'explain' this variation. We call these sources of variation 'Regression' and 'Residual'. It is a good idea to think of the Regression S.S. as the amount of variation in y which can be 'explained' by the other variable x, while the Residual S.S. is the remainder of the S.S. for y which is left unexplained. We use the six formulae, numbered (15.3) to (15.8), to calculate the sums of squares and degrees of freedom as follows:

$$\text{Total S.S.} = \Sigma y^2 - \frac{(\Sigma y)^2}{n} \tag{15.3}$$

$$\text{Regression S.S.} = b^2 \left[\Sigma x^2 - \frac{(\Sigma x)^2}{n} \right] \tag{15.4}$$

$$\text{Residual S.S.} = \text{Total S.S.} - \text{Regression S.S.} \tag{15.5}$$

$$\text{Total d.f.} = n - 1 \tag{15.6}$$

$$\text{Regression d.f.} = 1 \tag{15.7}$$

N.B. Formula (15.7) applies only when there is only one explanatory variable, x, say.

$$\text{Residual d.f.} = \text{Total d.f.} - \text{Regression d.f.} \tag{15.8}$$

For the example of the sales/test scores, and using the summations calculated earlier, we can now draw up the Analysis of Variance table:

Table 15.3. ANOVA Table for the Sales/Scores Data in Table 15.2

Source of Variation	S.S.	d.f.	M.S.	F Ratio
Regression	5338.4	1	5338.4	8.466
Residual	3783.5	6	630.6	
Total	9121.9	7		

Recall that the Mean Square (M.S.) column is obtained by dividing the S.S. values by the corresponding d.f., while the calculated F ratio is obtained as follows:

$$\text{Calc } F = \frac{\text{Regression M.S.}}{\text{Residual M.S.}} \text{ for 1 and } n - 2 \text{ d.f.} \tag{15.9}$$

For the example, $Calc \ F = \dfrac{5338.4}{630.6} = 8.47$ for (1,6) d.f., the d.f. values are those associated with the numerator and denominator of the ratio in the calculation of F.

This should by now seem like part of a hypothesis test, but what hypothesis is being tested? Remember that, in Chapter 10, the hypothesis tests specified a value of a population parameter such as the mean, or the difference between two means. What population are we talking about in regression analysis? The answer is the bivariate population of the two variables, x and y. Imagine that there is a population of points (one from each of a large number of salespeople) which could be plotted on our scatter diagram and that the regression line for this population is $y = \alpha + \beta x$, where α and β (Greek alpha and beta) are the intercept and slope of the population regression line. We calculate estimates a and b of these two population parameters based on our sample values of x and y. We are usually more interested in the slope rather than the intercept and, in fact, the ANOVA above helps us to test the null hypothesis that $\beta = 0$, which implies a horizontal regression line. Here are the formal steps of the hypothesis test:

1. $H_0 : \beta = 0$
2. $H_1 : \beta \neq 0$
3. 5% significance level
4. *Calc F* = 8.47 from the ANOVA table above
5. *Tab F* = 5.99, using Table C.6 for 1 and 6 d.f.
6. Since *Calc F* > *Tab F*, reject H_0 and conclude that:
7. 'The slope of the regression line of sales on test score is significantly different from zero (5% level)'.

Another way of expressing this conclusion is as follows: 'Test score (x) **explains** a significant amount of the variation in Sales (y)'.

I prefer the latter way of stating the conclusion, since it really refers to whether the variable x is useful as a predictor of y. Also it is possible to quantify the percentage of y which is **explained** by x, since this is given by:

$$\frac{\text{Regression S.S.}}{\text{Total S.S.}} \times 100\%$$

For the sales/test scores example, this is $\frac{5338}{9122} \times 100 = 58.5$. In other words: **58.5% of the variation in sales is explained by the test score**.

Before we discuss the assumptions of the F test above, there are three interesting points arising from the ANOVA table:

1. It can be shown that the square root of the ratio of the Regression S.S. to the Total S.S. is equal to Pearson's correlation coefficient, r, between the variables x and y. For our example,

$$r = \sqrt{\frac{5338}{9122}} = 0.765$$

2. The ANOVA table contains the value of the Residual S.S., i.e., 3783.5. We can compare this with the value of 3782 we obtained in Section 15.5. They agree, apart from rounding errors.

3. The Residual M.S. is a sample estimate of the population residual variance, which we will refer to as σ_r^2, while s_r^2 is our sample estimate. For the example, $s_r^2 = 630.6$ (see ANOVA table). So, $s_r = \sqrt{630.6} = 25.1$. We will be using this result later in this chapter.

However, we need to discuss briefly the assumptions of the F test which we carried out in the ANOVA. There are three assumptions, which are best understood by Fig. 15.2. Alternatively, in words, they refer to the distribution of the residuals as follows:

The distribution of the residuals is the same for all values of x (within the range of the sample data). It is approximately normal with a mean of zero and a standard deviation of σ_r^2.

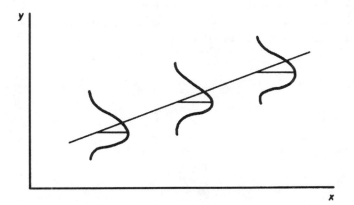

Figure 15.2   Assumptions Required in Using the Formula for Confidence Intervals for Predicted y

15.7   More Inferences in Regression Analysis

In this section we introduce two more useful applications of inference in Regression Analysis:

1. Confidence interval for β
2. Confidence interval for predicted values of y

Confidence Interval for the Slope of the Population Regression Line

The ANOVA we introduced earlier in this chapter was useful in many ways, one of which was a hypothesis test in which the null hypothesis was $H_0:\beta = 0$. It would be more interesting to have a confidence interval for β (rather than simply knowing that the slope was significantly different from zero). The appropriate formula is (15.10)

$$b \pm \frac{ts_r}{\sqrt{\Sigma x^2 - \frac{(\Sigma x)^2}{n}}} \qquad (15.10)$$

For our example,

1. $b = 1.904$,
2. For a 95% confidence interval, we obtain t from Table C.5 for $\alpha = 0.025$ and $\nu = n - 2 = 6$. So, in this case, $t = 2.447$.
3. $s_r = 25.1$ (see Section 15.6), so a 95% confidence interval for β is

$$1.904 \pm \frac{2.447 \times 25.1}{\sqrt{35,925 - \frac{525^2}{8}}}$$

$$1.904 \pm \frac{61.42}{\sqrt{1472}}$$

$$1.904 \pm 1.601$$

$$0.303 \text{ to } 3.505$$

Quite a wide spread, confirming how small samples can lead to results of little value, especially when data are very variable. If we wanted to know the slope to within ± 0.05, say, then we need to do calculations like those in Section 9.6.

Confidence Interval for Predicted Values of y

So far, our predictions for y have been single-value estimates for given values of x. The following Formula (15.11) enables us to calculate a confidence interval for a predicted value of y for any value of x, which we will refer to as x_0:

$$(a + bx_0) \pm ts_r \sqrt{\frac{1}{n} + \frac{(x_0 - \bar{x})^2}{\Sigma x^2 - \frac{(\Sigma x)^2}{n}}} \qquad (15.11)$$

The value of x_0 can be anywhere between 45 and 85 for our example. Let's try $x_0 = 60$. A 95% confidence interval for predicted y when $x = 60$, is

$$(9.4 + 1.904 \times 60) \pm 2.447 \times 25.1 \times \sqrt{\frac{1}{8} + \frac{(60 - 65.6)^2}{35925 - \frac{525^2}{8}}}$$

$$123.6 \pm 23.5$$

$$100.1 \quad \text{to} \quad 147.1$$

Looking again at the Formula (15.11), we can see that the 'error term' (which in our example is 23.5) depends on the value of x_0. It is smallest when $x_0 = \bar{x}$, which for our example is 65.6, but it increases if we choose values for x_0 away from the mean in either direction. For example, the error term is 39.5 when $x_0 = 45$, while it is 37.8 when $x_0 = 85$. The set of 95% confidence intervals for all values of x between 45 and 85 can be represented graphically by two curves (see Fig. 15.3.)

15.8 Regression Analysis Using Minitab for Windows

In this section, using Minitab for Windows we will outline the steps required to produce most of the results of the previous sections, which could all have been performed using only a hand calculator. We will, naturally, use the sales/test score example.

Start Minitab and when the 'data window' appears,

1. Type the sales data (from Table 15.1) into C1, rows 1–8.
2. Type the test score data (from Table 15.1) into C2, rows 1–8.

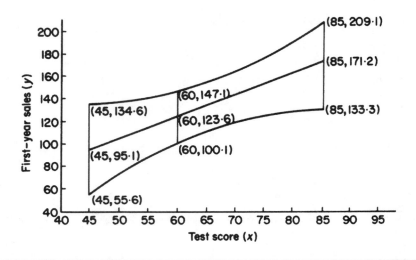

Figure 15.3 The Locus of 95% Confidence Intervals for Predicted _y_, Using Data from Table 15.1.

3. Think of some values of the '_x_ variable', i.e., test score, for which you require a single value prediction of the '_y_ variable', i.e., sales, and/or a 95% confidence interval for this prediction. For example we could choose 45, 60, and 85 (since we used those values earlier in this chapter).

Type these into C3, rows 1 to 3. Your data window should look like the first three columns of Table 15.3, assuming that you type in the names of the columns at the top of each column.

Now: Choose **Graph > Character Graph > Scatter Plot**
Enter **C1 in Y Variable** box
Click on **OK**
Enter **C2 in X Variable** box
Click on **OK**
Enter **C1** in **Response** box
Enter **C2** in **Predictor** box
Click in **Residuals** tiny box
Click in tiny box labelled **Fits**
Click on **Options**
Enter **C3** in **Predictor interval box**
Type **45, 60, 85** in box labelled **Predictor interval for new observation**

Table 15.4 Minitab Input Data and Some Output for the Sales/Test Score Example

Sales	Score	C3	Fits1	Res1
105	45	45	95.096	9.9045
120	75	60	152.229	−32.2293
160	85	85	171.274	−11.2739
155	65		133.185	21.8185
70	50		104.618	−34.6178
150	70		142.707	7.2930
185	80		161.752	23.2484
130	55		114.140	15.8599

Choose **File > display Data**
Type **C1 − C5** in box **'Columns and Constants to display'**
Click **OK**
Choose **File > Print Window**
The print-out contains a computer scatter diagram (Fig. 15.4). The rest of the output is referred to as Table 15.5.

Notes on the Computer Output (Fig. 15.4 and Table 15.5)

1. Minitab refers to the y variable as the **'Response'**. In our example, sales is the **Response**.
2. Minitab refers to the x variable as the **'Predictor'**. In our example, test score is the **Predictor**.
3. Much of Table 15.4 is self-explanatory. Just before the ANOVA table we see $s = 25.11$, (which we called s_r) which is the estimate of the residual standard deviation, σ. We also see that R-sq = 58.5%, which agrees with the number we obtained for the square of Pearson's r in section 15.6.
4. In the ANOVA table, Minitab uses the word 'Error' instead of 'Residual'. I prefer the word 'Residual', since it means that which remains to be explained.
5. The result that $p = 0.027$ implies that the null hypothesis, that the slope is zero, should be rejected at the 5% level of significance, since $0.027 < 0.05$.
6. After the ANOVA in Table 15.5, the estimates of y for x values of 45, 60, and 85 are given. For example, when $x = 45$, the single-value

Table 15.5 Further Output for the Sales/Score Regression Example

The regression equation is
Sales = 9.4 + 1.90 Score

Predictor	Coef	Stdev	t-rat	p
Constant	9.39	43.86	0.21	0.837
Score	1.9045	0.6545	2.91	0.027

s = 25.11   R-sq = 58.5%   R-sq(adj) =51.6%

Analysis of Variance

SOURCE	DF	SS	MS	F	p
Regression	1	5338.4	5338.4	8.47	0.027
Error	6	3783.4	630.6		
Total	7	9121.9			

Fit	Stdev.Fit	95% C.I.		95% P.I.	
95.10	16.16	55.55,	134.64	22.01,	168.18
123.66	9.61	100.14,	147.19	57.85,	189.47
171.27	15.48	133.38,	209.16	99.07,	243.48

ROW	Sales	Score	C3	FITS1	RESI1
1	105	45	45	95.096	9.9045
2	120	75	60	152.229	−32.2293
3	160	85	85	171.274	−11.2739
4	155	65		133.185	21.8153
5	70	50		104.618	−34.6178
6	150	70		142.707	7.2930
7	185	80		161.752	23.2484
8	130	55		114.140	15.8599

prediction of y is 95.10, while a 95% confidence interval for predicted y is 55.55 to 134.64.

7. The last block of output gives the input data (cols. 2-4) and the predicted values of y and the residuals for each data point.

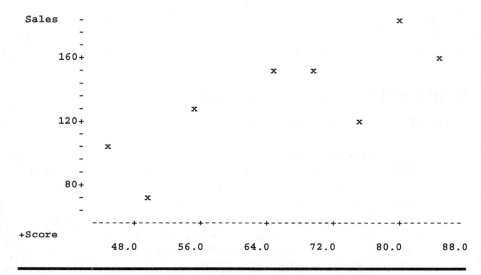

Figure 15.4   Minitab Output for Sales/Score Regression Example

You should check that the Residual S.S. = 3783.4 using two different methods.

15.9   Summary

Linear regression analysis is a method of deriving a linear equation relating two quantitative variables so that values of one of the variables, sometimes called the **Response** or y variable, can be predicted from the other variable sometimes called the **Explanatory** or **Predictor** or 'x', variable. The graphical representation of the regression equation is called the regression line. Formulae (15.1 and 15.2) give the intercept and slope of the line, which has the property that the sum of squares of the distances of the data points (on the scatter diagram of variables x and y) is smaller for this line than for any other line. These distances are called 'residuals'.

Inferences in regression analysis start with the use of ANOVA (previously introduced in Chapter 12) to test $H_0:\beta = 0$, where β is the slope of the (population) regression line. In this test, and in other inferential methods in regression analysis, we need to be able to assume that the residuals have the same normal distribution, i.e., with a mean of zero and a constant variance.

Two other inference examples in regression covered were (i) a confidence interval for β, and (ii) confidence intervals for the predicted values of y for all values of x, within the range of the sample data.

Worksheet 15:   Regression Analysis

Fill in the gaps in Questions 1 and 2.

1. The purpose of regression analysis is to .... values of one variable for particular values of another variable. We call the variable whose values we wish to predict the .... variable, and the other we call the .... variable.
2. Using sample values of the two variables the .... diagram may be drawn. If this appears to show a linear relationship between the variables we calculate a and b for the linear .... equation. This equation may be represented by a .... on the scatter diagram.

Questions 3, 4, and 5 are multiple choice. Choose one of the three options in each case.

3. In regression analysis, the y variable is chosen:
   (a) Arbitrarily,
   (b) As the variable plotted on the horizontal axis in the scatter diagram,
   (c) As the variable to be predicted.
4. The slope (or gradient) of a regression line:
   (a) Is always between -1 and $+1$,
   (b) Can have any value,
   (c) Can never be negative.
5. The purpose of calculating Pearson's r is to:
   (a) Replace points on a scatter diagram by a straight line,
   (b) Measure the degree to which two variables are linearly related.
   (c) Predict one variable from another variable.
6. The regression line is sometimes called the line of 'best fit' because it minimises the sum of squares of distances from the data points on the scatter diagram to the line in the y direction. For the example used in Chapter 15 (see Table 15.1 and Fig. 15.1), this (residual) sum of squares is 3783. Draw any other line 'by eye' on the scatter diagram which you think may be a better fit and calculate the residual sum of squares for your line. You should not be able to get below 3783, rounding errors excepted.

7. The following table gives the number of bathers at an open-air swimming pool and the maximum recorded temperature ($°C$) on ten Saturdays during one summer:

Number of Bathers	Maximum Temperature
290	19
340	23
360	20
410	24
350	21
420	26
330	20
450	25
350	22
400	29

Draw a scatter diagram and calculate the slope and intercept of the regression line which could be used to predict the numbers of bathers from the maximum temperature. Plot the regression line on the scatter diagram, checking that it passes through the point $(\bar{x}, \bar{y})$.

How many bathers would you predict if the forecast for the maximum temperature on the following Saturday in the summer was (a) 20°C, (b) 25°C, (c) 30°C? Which of the predictions will be the least reliable? Explain your reasons.

8. In order to estimate the depth of water (in metres) beneath the keel of a boat, a sonar measuring device was fitted. The device was tested by observing the sonar readings over a number of known depths, and the following data were collected:

Sonar reading	0.15	0.91	1.85	3.14	4.05	4.95
True depth of water	0.2	1.0	2.0	3.0	4.0	5.0

Draw a scatter diagram for these data and derive a linear regression equation which could be used to predict the true depth of water from the sonar reading. Predict the true depth from a sonar reading of zero and obtain a 95% confidence interval for your prediction. Interpret your result.

9. The percentage moisture content of a raw material and the percentage relative humidity of the atmosphere in the store where the

material as kept were measured on seven randomly selected days. On each day one randomly selected sample of material was used.

Relative humidity	30	35	52	38	40	34	60
Moisture	7	10	14	9	11	6	16

Draw a scatter diagram and derive a linear regression equation which could be used to predict the moisture content of the raw material from the relative humidity. Use the equation to predict moisture content for a relative humidity of (a) 0%, (b) 50%, (c) 100%. Also test the hypothesis that the slope of the population regression line is zero.

10. The data below give the weight (kg) and the daily food consumption (in hundreds of calories) for 12 obese adolescent girls. Calculate the best-fit linear regression equation which would enable you to predict food consumption from weight, checking initially that the relationship between the two variables appears to be linear.

Weight	85	95	80	60	95	85	90	80	85	70	65	75
Food consumption	32	33	33	24	39	32	34	28	33	27	26	29

What food consumption would you predict, with 95% confidence, for adolescent girls weighing (a) 65 kg, (b) 80 kg, (c) 95 kg?

11. To see if there is a relationship between the size of boulders (cm) in a stream and the distance (km) from the source of the stream, samples of boulders were measured at 1-km intervals. The average sizes of boulders found at various distances were as follows:

Distance downstream	1	2	3	4	5	6	7	8	9	10
Average boulder size	105	85	80	85	75	70	75	60	50	55

Obtain the regression equation which could be used to predict average boulder size from distance downstream. Plot the regression line on the scatter diagram. Test the null hypothesis that $\beta = 0$ against the alternative that $\beta \neq 0$. Also obtain a 95% confidence interval for β using Formula (15.10), where the value of t is from Table C.5 for $\alpha = 0.025$, and $\nu = n - 2$.

12. The number of grams of a given salt which will dissolve in 100 g of water at different temperatures (°C) is shown below:

Temperature	0	10	20	30	40	50	60	70
Weight		53.5	59.5	65.2	70.6	75.5	80.2	85.5	90.0

Obtain the regression equation which could be used to predict weight of salt from temperature. Plot the regression line on the scatter diagram. Predict the weight of salt which you estimate would dissolve at temperatures of (a) 25°C, (b) 55°C, (c) 85°C. Comment on your results.

13. A random sample of ten people who regularly attempted the daily crossword puzzle in a certain national newspaper were asked to time themselves on a puzzle which none of them had seen before. Their times (in minutes) to complete the puzzle and their scores in a standard IQ test were as follows:

IQ	120	100	130	110	100	140	130	110	150	90
Times	9	7	13	8	4	5	16	7	5	13

What conclusions can be drawn from these data? Why is drawing a 'line by eye' through the data for this question much easier than compared with the time regression line for Question 12?

Chapter 16

Goodness-of-Fit Tests

16.1 Introduction

We return, in this the final chapter, to a one-variable problem, namely, the problem of deciding whether our sample data could have been selected from a particular type of probability distribution. Four types of distribution will be considered:

Type of Distribution	Type of Variable
'Simple proportion'	Categorical
Binomial	Discrete
Poisson	Discrete
Normal	Continuous

For the first three types, a χ^2 test will be used to see how closely the observed frequencies of the sample data agree with the frequencies we would expect under the null hypothesis that the sample data actually do come from the type of distribution being considered (refer to Chapter 13 if you are unfamiliar with the χ^2 test). For the fourth type, namely, the normal distribution, a small-sample test, called the Shapiro-Wilk test, is discussed.

16.2 Goodness-of-Fit for a Simple Proportion Distribution

We define a simple proportion distribution as one for which the expected frequencies of the various categories, into which the 'values' of a categorical variable will fall are in certain numerical proportions or ratios.

Example

In a standard pack of 52 playing cards, the ratio of the numbered cards (2, 3,...,10) to unnumbered cards (Ace, King, Queen, Jack) is 36 to 16, which we could write as 36:16. Suppose we selected cards randomly with replacement. Then we would expect the proportion of numbered cards to be 36/52, while the proportion of unnumbered cards expected would be 16/52.

Another Example

Suppose that there is a genetic theory that adults should have hair colours of black, brown, fair, and red in the ratios 5:3:1:1. If this theory is correct, we would expect the frequencies of black, brown, fair, and red hair to be in the proportions:

$$\frac{5}{5+3+1+1}, \quad \frac{3}{5+3+1+1}, \quad \frac{1}{5+3+1+1}, \quad \frac{1}{5+3+1+1},$$

or $\dfrac{5}{10}, \dfrac{3}{10}, \dfrac{1}{10}, \dfrac{1}{10}$,

or even simpler: 5:3:1:1.

If we take a random sample of 50 people to test this theory, we would expect 25 to have black hair, 15 to have brown hair, 5 to have fair hair, and 5 to have red hair. We simply multiplied the expected proportions by the total sample size of 50. Now we compare these expected frequencies with the frequencies we actually observed in the sample, and calculate an χ^2 statistic.

It is convenient to set out this calculation in the form of a table (Table 16.1). Notice that the method for calculating the E values ensures that the sum of the E values equals the sum of the O values.

We now set out the seven steps in the hypothesis test for this example:

1. H_0: Sample data support the Genetic theory of a 5:3:1:1 distribution.
2. H_1: Sample data do not support the Genetic theory.
3. 5% significance level.
4. From Table 16.1, we have

$$Calc \ \chi^2 = \sum \frac{(O-E)^2}{E} = 1.36$$

Table 16.1   Calculation of χ^2 for a 5:3:1:1 Distribution

Hair Colour	Expected Proportions	Expected Frequencies (E)	Observed Frequencies (O)	$\dfrac{(O-E)^2}{E}$
Black	5/10	25	28	0.36
Brown	3/10	15	12	0.60
Fair	1/10	5	6	0.20
Red	1/10	5	4	0.20
		50	50	Calc χ^2 = 1.36

Table 16.2   Results of an ESP Experiment, 50 Subjects, 5 Trials per Subject

Number of correct decisions	0	1	2	3	4	5
Number of subjects	15	18	8	5	3	1

5. *Tab* χ^2 = 7.82 (even though H_1 is two-sided), and ν = (number of categories − 1) = 4 − 1 = 3, from Table C.10.
6. Since *Calc* χ^2 < *Tab* χ^2, do not reject the null hypothesis.
7. It is reasonable to assume that the Genetic theory is correct, from the evidence provided by the sample data given in Table 16.1.

Notes

(a) The formula for *Calc* χ^2 is only valid if all the E values are at least 5. If any E value is less than 5, it may be sensible to combine adjacent categories so that all E values for the new categories are all at least 5.
(b) The formula

$$Calc\ \chi^2 = \sum \frac{(O-E)^2}{E} \tag{13.3}$$

is used if $\nu > 1$. If $\nu = 1$, use

$$Calc\ \chi^2 = \sum \frac{(|O-E| - \frac{1}{2})^2}{E} \tag{13.1}$$

(applying 'Yates's correction' as in Section 13.3).

(c) The formula for degrees of freedom $\nu = $ (number of categories $- 1$) may be justified by reference to Section 9.7, the one restriction being that the sum of the E values must be made equal to the sum of the O values. In our example, only three of the E values may be determined independently, so there are three degrees of freedom.

(d) If some of the categories are combined to avoid low E values, the number of categories *after combinations* is used in the formula for degrees of freedom.

16.3 Goodness-of-Fit for a Binomial Distribution

Suppose we carry out n Bernoulli trials in which each trial can result in two possible outcomes, which we call 'success' and 'failure'. The trials are independent and the probability of success is constant. Suppose we repeat this set of n trials several times and observe the frequencies of the number of successes that occur. We may then carry out a χ^2 test to decide whether it is reasonable to conclude that the number of successes in n trials has a binomial distribution, with a value of p which we can either estimate from the observed frequencies or sometimes specify without reference to the observed frequencies. (It will be assumed that you are familiar with the Bernoulli and the binomial distributions as described in Chapter 6.)

Example

In an experiment in extrasensory perception (ESP), four cards marked either A, B, C, or D were used. The experimenter, unseen by the subject, shuffles the cards and selects one. The subject tries to decide which card has been selected, and having decided writes down A, B, C, or D. This procedure is repeated five times for each of a random sample of 50 subjects. The number of times, out of a maximum of five, that each subject correctly identifies a selected card is counted. Suppose that the data for all 50 subjects are recorded in a table such as Table 16.2. Is there evidence that subjects are simply guessing, or do the subjects have powers of ESP, which would presumably have significantly more 'correct decisions'?

We can regard the testing of each subject as a set of five Bernoulli trials, each trial having one of two possible outcomes, 'correct decision' or 'incorrect decision'. This set of trials is repeated (on different subjects) a total of 50 times. The second row of Table 16.2 gives the observed frequencies (O) for the various possible numbers of correct decisions, for each subject.

If subjects are guessing, then the probability of a correct decision each time a subject guesses is $\frac{1}{4}$ or 0.25, since the four types of card are equally likely to be selected. So the question above, 'Is there evidence that the

Table 16.3 Calculation of χ^2 for a Binomial Distribution

Number of correct decisions	P(x)	E = P(x) × 50	O	$\dfrac{(O - E)^2}{E}$
0	0.2373	11.9	15	0.81
1	0.3955	19.8	18	0.16
2	0.2637	13.2	8	2.05
3	0.0879	4.4 ⎤	5 ⎤	
4	0.0146	0.7 ⎬ 5.2	3 ⎬ 9	2.78
5	0.0010	0.1 ⎦	1 ⎦	
Total	1.0000	50.1	50	Calc χ^2 = 5.80

Notes:

(a) The probabilities P(x) were obtained from Table C.1 for $n = 5$ and $p = 0.25$.
(b) The bottom three categories in Table 16.3 have been combined to ensure that all E values, after combinations, are at least 5. In the example, there were 6 categories initially, but this has reduced to 4 after combinations (see the final column of Table 16.3).
(c) The totals of the E and the O columns are equal (apart from rounding errors).

subjects are simply guessing?' is equivalent to the question: 'Is it reasonable to suppose that the data in Table 16.2 come from a binomial distribution with $n = 5$ and $p = 0.25$?'

The expected frequencies (E) for the various numbers of correct are obtained by assuming, for the purposes of the calculation, that we are dealing with a B(5, 0.25) distribution. First we calculate the probabilities of 0, 1, 2, 3, 4, and 5 correct decisions (using the methods of Chapter 6). These probabilities are multiplied by the total of the observed frequencies (50 in our example) to give the expected frequencies. These calculations and the calculation of χ^2 are set out in Table 16.3

We now set out the seven-step hypothesis test for this example:

1. H_0: Sample data do come from a B(5, 0.25) distribution, implying that the subjects are guessing.
2. H_1: Sample data do not come from a B(5, 0.25) distribution, which might imply that some subjects have powers of ESP.
3. 5% significance level.
4. Calc $\chi^2 = 5.80$ from Table 16.3.
5. Tab $\chi^2 = 7.82$ for $\alpha = 0.05$, $\nu = $ number of categories $- 1 = 3$. (Refer to notes (a) and (b) below and Table C.9.)
6. Since Calc $\chi^2 <$ Tab χ^2, do not reject H_0
7. It is reasonable to suppose that subjects are guessing (5% level).

16.4   Goodness-of-Fit for a Poisson Distribution

Suppose that we observe the number of times a particular event occurs in each of a number of units of time (or space). Can we conclude that the number of occurrences of the event per unit time (or space) has a Poisson distribution, implying randomly occurring events?

Example

Suppose that the number of major earthquakes occurring per month in a particular geographical area is collected for 100 months and summarised, as in Table 16.4.

The observed frequencies (O) for the various numbers of earthquakes per month are given in the second row of the Table 16.4. The expected frequencies (E) for the various numbers of earthquakes are obtained by assuming, for the purposes of the calculation, that we are dealing with a Poisson distribution. The parameter m, the mean of the distribution, is 'estimated from the data' by calculating the sample mean number of earthquakes per month using the following 'common sense' formula (not on the formula list):

$$m = \frac{\text{total number of earthquakes}}{\text{total number of months}}$$

$$= \frac{57 \times 0 + 31 \times 1 + 8 \times 2 + 3 \times 3 + 1 \times 4}{100}$$

$$= 0.6$$

Note

In cases where the value of p is not specified by the experimental set-up, unlike the example above, we must estimate p from the data of the observed frequencies (see Worksheet 16, Question 7), and we 'lose' a further degree of freedom.

Table 16.4   Number of Earthquakes Occurring in 100 Months

Number of earthquakes per month	0	1	2	3	4
Number of months	57	31	8	3	1

Table 16.5 Calculation of χ^2 for a Poisson distribution

Number of Earthquakes per Month (x)	P(x)	E = P(x) × 100	O	$\frac{(O-E)^2}{E}$
0	0.5488	54.9	57	0.08
1	0.3293	32.9	31	0.11
2	0.0988	9.9 ⎫	8 ⎫	
3	0.0197	2.0 ⎬ 12.2	3 ⎬ 12	0.00
4 or more	0.0034	0.3 ⎭	1 ⎭	
Total	1.0000	100.0	100	Calc χ^2 = 0.10

Notes:

(a) The probabilities P(x) were obtained from Table C.2 for $m = 0.6$ (see Section 6.12).

(b) The probability of '4 or more' (rather than '4') is calculated to ensure that the totals of the E and O columns are equal, apart from rounding errors.

(c) The bottom three categories in Table 16.5 have been combined to ensure that, after combinations, all expected values are at least 5.

For the Poisson distribution with a mean $m = 0.6$, we can obtain the probabilities of 0, 1, 2, 3, and 4 or more earthquakes (using the methods of Chapter 6). These probabilities are multiplied by the total of the observed frequencies (100 in the example) to give the expected frequencies

These calculations and the subsequent χ^2 test statistic are set out in Table 16.5. We now give the seven-step hypothesis test for this example:

1. H_0: Sample data come from a Poisson distribution, implying that earthquakes occur randomly in time.
2. H_1: Sample data do not come from a Poisson distribution.
3. 5% significance level.
4. Calc $\chi^2 = 0.19$ from Table 16.5.
5. Tab $\chi^2 = 3.48$ for $\alpha = 0.05$, ν = number of categories $-1 - 1 = 1$ (see footnote (a) below).
6. Since Calc $\chi^2 <$ Tab χ^2, do not reject H_0.
7. It is reasonable to assume a Poisson distribution, and that earthquakes occur randomly in time (5% level).

Notes

There are only three categories after combinations. One degree of freedom is lost because of the restriction $\Sigma E = \Sigma O$, and another is lost because the parameter m has been estimated from the sample data.

Strictly speaking we should have used Yates' correction in calculating χ^2 because this has become a 1 d.f. example (see step 4). However, this would have had the effect of reducing the value of *Calc* χ^2, and so the null hypothesis would still not have been rejected.

16.5   The Shapiro–Wilk Test for Normality

Although it is possible to use a χ^2 test for normality if we have a sample of at least 50, the assumption of normality required to carry out most hypothesis tests: (a) Only requires the assumption of *approximate* normality, (b) Is less important if the sample size is large.

The problem of normality is really only practically important for sample sizes below about $n = 15$, as we pointed out in Section 9.4. There we used dotplots and judgement. A more objective method for small samples is provided by the Shapiro-Wilk test as follows.

Rank the n sample observations in increasing order, referring to them as $x_{(1)}, x_{(2)}, \ldots, x_{(n-1)}, x_{(n)}$, where $x_{(1)}$ refers to the smallest observed value, $x_{(2)}$ the next smallest, and so on, and finally $x_{(n)}$ is the largest.

Then calculate:

$$b = a_1 (x_{(n)} - x_{(1)}) + a_2(x_{(n-1)} - x_{(2)}) + \ldots \qquad (16.1)$$

where $a_1, a_2, \ldots$ are coefficients taken from Table C.13. Now calculate the test statistic *Calc W*, given by Formula (16.2):

$$Calc\ W = \frac{b^2}{(n-1)s^2} \qquad (16.2)$$

where s is, of course, the standard deviation of the n observations, given by Formula 4.4, for example. The next step is to look up *Tab W* in Table C.14. The value of *Tab W* depends on n and the significance level, which we usually take to be 5% (i.e., 0.05). *Calc W* > *Tab W*, we do not reject the null hypothesis of normality, and hence it is reasonable to assume that our data do come from a normal distribution.

Example (in which n is an even number)

Test the normality of the following sample of 10 observations (data from Section 11.2).

    200   550   290   170   180   350   190   210   160   250

In rank order these are

    160   170   180   190   200   210   250   290   350   550

Introducing the coefficients from Table C.13:

b $= 0.5739(550 - 160)\ + 0.3291(350 - 170)\ + 0.2141(290 - 180)$

    $+\ 0.1224(250 - 190) + 0.0399(210 - 200)$

   $=\ 223.8 + 59.24 + 23.55 + 7.34 + 0.40$

   $=\ 314.35$

Since $s = 119.47$ for these data,

$$Calc\ W = \frac{314.35^2}{(10 - 1)119.47^2} = 0.791$$

Table C.14 gives *Tab W* $= 0.842$ for $n = 10$ and a 5% significance level. Since *Calc W* $<$ *Tab W*, $(0.769 < 0.842)$ reject normality (as we thought we should do in Section 11.2, basing our conclusions mainly on a dotplot which indicated positive skewness).

Example (n is an odd number this time)

Test the normality of the following sample of 9 observations (data taken from Section 9.4):

    163   157   160   168   155   168   164   157   169

In rank order these are

    155   157   157   160   163   164   168   168   169

Introducing the coefficients from Table C.13:

$b = 0.5888(169 - 155)\ + 0.3244(168 - 157) + 0.1976(168 - 157)$

    $+\ 0.0947(164 - 160)$

   $=\ 8.24 + 3.57 + 2.17 + 0.38$

   $=\ 14.36$

Since $s = 5.34$ for these data,

$$Calc\ W = \frac{14.36^2}{(9 - 1)5.34^2} = 0.904$$

Tab W $= 0.829$, and hence *Calc W* $>$ *Tab W*, so we do not reject normality.

We conclude that these data are consistent with a normal distribution (as we thought was the case in Section 9.4), basing our decision on a simple dotplot [Fig. 9.5], which indicated that the distribution appeared to be fairly symmetrical, although not bunched in the middle. It was, to the eye, similar to the dotplots in Fig. 9.4, which were based on samples from a normal distribution.

16.6   Summary

Goodness-of-fit tests are tests to decide whether it is reasonable to conclude that a sample of univariate (one-variable) data could have been drawn from a particular type of distribution. Four types of distribution were covered, namely, the simple proportion, binomial, Poisson, and the normal. The first three were tested using a χ^2 test, the last by means of the Shapiro-Wilk test.

For the χ^2 test, the sample data are in the form of observed frequencies. Expected frequencies are calculated assuming that the sample data do come from the particular distribution under investigation. The degrees of freedom for *Tab* χ^2 are, in general, equal to:

(number of categories after combinations) − (number of distribution

parameters estimated from the sample data) − 1.

Worksheet 16:   Goodness-of-Fit Tests

1. In Mendel's experiments with peas, he classified each of 556 peas into one of four categories as follows:

Type of Pea	Number of Peas
Round and yellow	315
Round and green	108
Wrinkled and yellow	101
Wrinkled and green	32

Are these data consistent with Mendel's theory of heredity that these categories should occur in the proportions 9:3:3:1?

2. The number of fatal road accidents in one year in a large city were tabulated according to the time they occurred:

Time	Midnight to 4 a.m.	4 a.m. to 8 a.m.	8 a.m. to noon	noon to 4 p.m.	4 p.m. to 8 p.m.	8 p.m. to midnight
Number of accidens	28	15	14	18	15	30

Test the hypotheses that:

(a) Accidents are uniformly distributed in time.

(b) Accidents occur in the ratios $2 : 1 : 5 : 4 : 5 : 3$, these being the estimated ratios of the volumes of traffic occurring in the city for the six four-hour periods.

3. The number of sheep farms of a given size in a county and the type of land on which they are situated were as follows:

Type of Land	Number of Sheep Farms
Flat	43
Hilly	32
Mountainous	5

If 35% of the county is flat, 50% is hilly, and 15% is mountainous, is the number of farms independent of the type of land?

4. A random sample of 100 families were asked how many cars they owned. The results were

Number of cars	0	1	2 or more
Number of families	35	45	20

Test the hypothesis that, for all families, the ratios are $1:2:1$ for the three categories of the number of cars owned.

5. For a random sample of 300 families each with three children, the distribution of the number of boys was as follows:

Number of boys	0	1	2	3
Number of families	55	108	102	35

Test the hypothesis that the number of boys in families with three children has:

(a) A binomial distribution with $p = 0.5$, implying that boys and girls are equally likely at each birth.

(b) A binomial distribution. *Hint*: estimate p from the sample data using the relative frequency definition, i.e.,

$$\frac{\text{total number of boys in the 300 families}}{\text{total number of children in the 300 families}}$$

Compare the conclusions of (a) and (b).

6. Samples of 10 pebbles were taken from each of 200 randomly selected sites on a beach. The number of limestone pebbles in each sample was counted. The results are summarised in the following table:

Number of limestone pebbles	0	1	2	3	4	5	6	7	8	9	10
Number of sites	0	7	20	45	53	39	25	8	3	0	0

How would you have selected the sites? Is it reasonable to conclude that the number of limestone pebbles in samples of 10 has a binomial distribution with a parameter, $p = 0.4$?

7. An experiment was carried out to test whether the digit '8' occurred randomly in random number tables. Successive sets of 20 single digits (0, 1, 2, ..., 9) were examined and the number of times the digit '8' occurred was noted for each set.

Number of '8' digits found	0	1	2	3	4	5	6 or more
Number of sets	25	45	70	35	15	10	0

What conclusion can be drawn from these data?

8. A survey was conducted to decide whether a particular plant species was randomly distributed in a meadow. Eighty points in the meadow were randomly selected. A quadrat was placed with its centre at each of the selected points and the number of individual plants of the species was noted:

Number of plants per quadrat.	0	1	2	3	4	5	6
Number of quadrats	11	37	12	7	6	4	3

How would you have selected 80 points randomly in a meadow? Is it reasonable to conclude that the plant species was randomly distributed in the meadow?

9. The number of dust particles occurring in unit volumes of gas was counted. The procedure was repeated 100 times for the same constant volume. Given the following results, is it reasonable to assume that the number of dust particles per unit volume is randomly distributed with a mean of two particles per unit volume?

Number of particles	0	1	2	3	4 or more
Number of times this number of particles observed	9	32	26	15	18

10. The number of minor defects noted by an inspector in 90 cars leaving a production assembly line was as follows:

Number of defects	0	1	2	3	4	5	6
Number of cars	35	13	6	5	18	10	3

(a) Test whether the mean and variance of the number of defects per car are approximately equal. (This is a quick but not very reliable test for a Poisson distribution.)

(b) Now use the χ^2 test to decide whether the number of defects per car is randomly distributed.

11. Test the following data sets for normality:
(a) The ten differences listed in Table 9.2.
(b) The A-level counts of the nine students listed in Table 1.1.

Appendix A

Statistical Formulae

Sample mean $\bar{x} = \dfrac{\Sigma x}{n}$ (2.1)

Σx^2 means square the n observed values of x and then sum   (2.2)

$(\Sigma x)^2$ means sum the n observed values of x and then square this sum

      (2.3)

$\Sigma(x - \bar{x})$ means subtract the sample mean from each observed value of x and then sum   (2.4)

$n! = 1 \times 2 \times 3 \times \ldots \times n$ (n must be a positive integer)   (2.5)

Sample median is the $(n + 1)/2^{\text{th}}$ value   (4.1)

Sample mode is the value with the highest frequency   (4.2)

Sample standard deviation $s = \sqrt{\dfrac{\Sigma(x - \bar{x})^2}{n - 1}}$ (4.3)

Sample standard deviation $\quad s = \sqrt{\dfrac{\Sigma x^2 - \dfrac{(\Sigma x)^2}{n}}{n - 1}}$ (4.4)

Sample lower quartile, Q1, is the $(n + 1)/4^{\text{th}}$ value (4.5)

Sample upper quartile, Q3, is the $3(n + 1)/4^{\text{th}}$ value (4.6)

Inter-quartile range $=$ upper quartile $-$ lower quartile $=$ Q3 $-$ Q1

(4.7)

Measure of skewness $= \dfrac{3(\text{sample mean} - \text{sample median})}{\text{standard deviation}}$ (4.8)

Probability, the *a priori* definition:

$P(E) = \dfrac{r}{n} \quad$ where r out of n equally likely outcomes result in event E

(5.1)

Probability, the relative frequency definition:-

$P(E) = \dfrac{r}{n} \quad$ where r out of n trials results in event E, and n is large

(5.2)

Multiplication law (general case)

$$P(E_1 \text{ and } E_2) = P(E_1)P(E_2|E_1)$$ (5.3)

Multiplication law (special case)
If E_1 and E_2 are independent events,

$$P(E_1 \text{ and } E_2) = P(E_1)P(E_2)$$ (5.4)

Addition law (general case)

$$P(E_1 \text{ or } E_2 \text{ or both}) = P(E_1) + P(E_2) - P(E_1 \text{ and } E_2)$$ (5.5)

Addition law (special case)

If E_1 and E_2 are mutually exclusive events,

$$P(E_1 \text{ or } E_2) = P(E_1) + P(E_2) \tag{5.6}$$

Complementary events, a useful result

$$P(\text{at least } 1 \cdots) = 1 - P(\text{none} \cdots) \tag{5.7}$$

Bernoulli distribution; probability function $P(x)$, where

$$P(x) = p^x (1 - p)^{1 - x}, \qquad x = 0, 1 \tag{6.1}$$

Bernoulli distribution; mean and standard deviation:

$$\text{mean} = p \qquad \text{standard deviation} = \sqrt{p(1 - p)} \tag{6.2}$$

Binomial distribution; probability function $P(x)$, where

$$P(x) = \binom{n}{x} p^x (1 - p)^{n - x}, \qquad x = 0, 1,..., n \tag{6.3}$$

Binomial distribution; mean and standard deviation:

$$\text{mean} = np \qquad \text{standard deviation} = \sqrt{np(1 - p)} \tag{6.4}$$

Poisson distribution; probability function $P(x)$, where

$$P(x) = \frac{e^{-m} m^x}{x!}, \quad x = 0, 1,... \tag{6.5}$$

Poisson distribution; mean and standard deviation:

$$\text{mean} = m \qquad \text{standard deviation} = \sqrt{m} \tag{6.6}$$

Geometric distribution; probability function $P(x)$, where

$$P(x) = (1 - p)^{x - 1} p, \quad x = 1, 2,... \qquad \text{(Worksheet 6, Question 28)}$$

Standardization formula for normal distribution

$$z = \frac{(x - \mu)}{\sigma} \qquad (7.1)$$

Sampling distribution of the sample mean:

$$\mu_{\bar{x}} = \mu \qquad \sigma_{\bar{x}} = \frac{\sigma}{\sqrt{n}} \qquad (8.1,\ 8.2)$$

95% confidence interval for the mean, μ, of a population, large sample size, n

$$\bar{x} \pm \frac{1.96s}{\sqrt{n}} \qquad (9.1)$$

Confidence interval for the mean, μ, of a population, small sample size, n

$$\bar{x} \pm \frac{ts}{\sqrt{n}} \qquad (9.2)$$

95% confidence interval for a binomial probability

$$\frac{x}{n} \pm 1.96 \sqrt{\frac{\frac{x}{n}(1 - \frac{x}{n})}{n}} \qquad (9.3)$$

Confidence interval for the mean of a population of differences, paired samples data

$$\bar{d} \pm \frac{ts_d}{\sqrt{n}} \qquad (9.4)$$

where the mean and standard deviation of the differences are given by:

$$\bar{d} = \frac{\Sigma d}{n} \qquad s_d = \sqrt{\frac{\Sigma d^2 - \frac{(\Sigma d)^2}{n}}{n - 1}} \qquad (9.5)$$

Confidence interval for the difference in the means of two populations, unpaired samples data.

$$(\bar{x}_1 - \bar{x}_2) \pm ts\sqrt{\frac{1}{n_1} + \frac{1}{n_2}} \tag{9.6}$$

Pooled estimate, s^2, of the common variance of two unpaired populations

$$s^2 = \frac{(n_1 - 1)s_1^2 + (n_2 - 1)s_2^2}{n_1 + n_2 - 2} \tag{9.7}$$

Test statistic for testing the mean, μ, of a population

$$Calc \; t = \frac{\bar{x} - \mu}{\frac{s}{\sqrt{n}}} \tag{10.1}$$

Test statistic for testing a binomial probability

$$Calc \; z = \frac{\frac{x}{n} - p}{\sqrt{\frac{p(1 - p)}{n}}} \tag{10.2}$$

Test statistic for testing a population of differences

$$Calc \; t = \frac{\bar{d}}{\frac{s_d}{\sqrt{n}}} \tag{10.3}$$

Test statistic for testing the difference in the means of two populations

$$Calc \; t = \frac{\bar{x}_1 - \bar{x}_2}{s\sqrt{\frac{1}{n_1} + \frac{1}{n_2}}} \tag{10.4}$$

Test statistic for testing the equality of two variances

$$Calc \; F = \frac{s_1^2}{s_2^2} \quad \text{if} \quad s_1^2 > s_2^2 \quad \text{or} \quad Calc \; F = \frac{s_2^2}{s_1^2} \quad \text{if} \quad s_1 < s_2^2 \tag{10.5}$$

Normal approximation to sign test

$$\mu = \frac{n}{2} \qquad \sigma = \frac{\sqrt{n}}{2} \tag{11.1}$$

Normal approximation to Wilcoxon signed rank test

$$\mu_T = \frac{n(n+1)}{4} \qquad \sigma_T = \sqrt{\frac{n(n+1)(2n+1)}{24}} \tag{11.2}$$

Mann-Whitney U test

$$U_1 = n_1 n_2 + \frac{1}{2} n_1 (n_1 + 1) - R_1 \tag{11.3a}$$

$$U_2 = n_1 n_2 + \frac{1}{2} n_2 (n_2 + 1) - R_2 \tag{11.3b}$$

Normal approximation to Mann-Whitney U test

$$\mu_U = \frac{n_1 n_2}{2} \qquad \sigma_U = \sqrt{\frac{n_1 n_2 (n_1 + n_2 + 1)}{12}} \tag{11.4}$$

One-way ANOVA to compare two means

$$\text{Total S.S.} = \Sigma x_1^2 + \Sigma x_2^2 - \frac{G^2}{n_1 + n_2} \tag{12.1}$$
$$\text{where } G = \Sigma x_1 + \Sigma x_2$$

$$\text{Between S.S.} = \frac{(\Sigma x_1)^2}{n_1} + \frac{(\Sigma x_2)^2}{n_2} - \frac{G^2}{n_1 + n_2} \tag{12.2}$$

$$\text{Within S.S.} = \text{Total S.S.} - \text{Between S.S.} \tag{12.3}$$

$$\text{Total d.f.} = n_1 + n_2 - 1 \tag{12.4}$$

$$\text{Between d.f.} = 2 - 1 = 1 \tag{12.5}$$

$$\text{Within d.f.} = \text{Total d.f.} - \text{Between d.f.} \tag{12.6}$$

$$Calc\ F = \frac{\text{Between M.S.}}{\text{Within M.S.}} \quad \text{for} \quad 1 \quad \text{and} \quad n_1 + n_2 - 2 \quad \text{d.f.} \tag{12.7}$$

Posterior test following one-way ANOVA

$$\text{s.e.} = \sqrt{\frac{\text{Residual M.S.}}{n_{ij}}} \tag{12.8}$$

For each pair of treatments, i and j, compare the difference in the treatment means/s.e. with critical values of the Studentized range statistic, q, from Table C.9.

χ^2 statistic for a 2 × 2 contingency table:

$$Calc\ \chi^2 = \sum \frac{\left(|O - E| - \frac{1}{2}\right)^2}{E} \quad \text{if } v = 1 \tag{13.1}$$

$$E = \frac{\text{row total} \times \text{column total}}{\text{grand total}} \tag{13.2}$$

χ^2 statistic for a $r \times c$ contingency table:

$$Calc\ \chi^2 = \sum \frac{(O - E)^2}{E} \quad v = (r - 1)(c - 1) \tag{13.3}$$

Fisher exact test:

$$
\begin{array}{cc|c}
a & b & a + b \\
c & d & c + d \\
\hline
a + c & b + d & n = a + b + c + d
\end{array}
$$

$$\text{Probability} = \frac{(a + b)!(c + d)!(a + c)!(b + d)!}{n!a!b!c!d!}$$

The formula gives the probability of observing a, b, c, and d, while keeping the marginal row and column totals constant.

Correlation

$$r = \frac{\sum xy - \frac{\sum x \sum y}{n}}{\sqrt{\left[\sum x^2 - \frac{(\sum x)^2}{n}\right]\left[\sum y^2 - \frac{(\sum y)^2}{n}\right]}} \tag{14.1}$$

$$Calc\ t\ =\ r\sqrt{\frac{n-2}{1-r^2}} \tag{14.2}$$

$$r_s\ =\ 1-\frac{6\Sigma d^2}{n^3-n} \tag{14.3}$$

Regression Analysis

Linear regression equation of y on x: $y = a + bx$ (15.1)

$$b\ =\ \frac{\Sigma xy - \frac{\Sigma x \Sigma y}{n}}{\Sigma x^2 - \frac{(\Sigma x)^2}{n}} \qquad a\ =\ \bar{y} - b\bar{x} \tag{15.2}$$

$$\text{Total S.S.}\ =\ \Sigma y^2 - \frac{(\Sigma y)^2}{n} \tag{15.3}$$

$$\text{Regression S.S.}\ =\ b^2\left[\Sigma x^2 - \frac{(\Sigma x)^2}{n}\right] \tag{15.4}$$

Residual S.S. = Total S.S. − Regression S.S. (15.5)

Total d.f. = n − 1 (15.6)

Regression d.f. = 1   N.B. this applies only to the case where there is only one explanatory variable, x, say (15.7)

Residual d.f. = Total d.f. − Regression d.f. (15.8)

$$Calc\ F\ =\ \frac{\text{Regression M.S.}}{\text{Residual M.S.}}\quad \text{for 1 and } n-2 \text{ d.f.} \tag{15.9}$$

Confidence interval for β in regression analysis

$$b \pm \frac{ts_r}{\sqrt{\Sigma x^2 - \frac{(\Sigma x)^2}{n}}} \qquad (15.10)$$

Confidence interval for a predicted value of y in regression analysis

$$a + bx_0 \pm ts_r \sqrt{\frac{1}{n} + \frac{(x_0 - \bar{x})^2}{\Sigma x^2 - \frac{(\Sigma x)^2}{n}}} \qquad (15.11)$$

Shapiro-Wilk test

$$b = a_1(x_{(n)} - x_{(1)}) + a_2(x_{(n-1)} - x_{(2)}) + \cdots \qquad (16.1)$$

$$Calc\ W = \frac{b^2}{(n-1)s^2} \qquad (16.2)$$

Appendix B

Solutions to Worksheets

Worksheet 1 (Solutions)

1. (a)
2. (a) discrete; 0–200 hours; lightbulb.
   (b) continuous; £40–£100; hotel.
   (c) categorical; professional, skilled, ... adult male.
   (d) discrete; 0, 1, 2, ...10; 100 hours.
   (e) categorical: 0–100 hours; 100 hours.
   (f) discrete; 0, 1,.....100; month.
   (g) ranked; 1st, 2nd,.......25th; annual contest.
   (h) continuous; 0–300 cm; county.
   (i) discrete; 0, 1,.... 20; year (in period 1900–1999).
   (j) discrete; 0, 1,.....10; series of 10 encounters with T-junction.
   (k) ranked; A, B, C, D, E, N, U; candidate.
   (l) categorical; black, brown, ...; person.
   (m) categorical; presence, absence; sq. metre of meadow.
   (n) continuous; 0–5 seconds; rat.
   (o) continuous; 0–10 kg; tomato plant.
   (p) categorical; sand, limestone, ...; core sample.
   (q) categorical; Conservative, Labour, Liberal Democrats, ... ;
       person voting.
3. Age 0–3 years could be category 'pre-school';
   similarly 'nursery school' for 3–5 yrs,
   'primary school' for 5–11 yrs,
   'secondary school' for 11–16 yrs,

'sixth-form college' for 16–18 yrs,
'university' for 18–21 yrs,
'employment' for 21–60 (women) or 65 yrs (men).
'retirement' for 60 or 65 yrs and over.
The logical order is as above, i.e., according to increasing age.

4. (a) Yes, because although the values taken by the ranked variable may appear to be numerical, for example 1st, 2nd, 3rd, ... in a race, the variable does not have values 1, 2, 3, and so on. A, B, C, would do just as well.

5. (a) (i) According to a book called *An Introduction to the Theory of Statistics* by Yule, G. U. and Kendall, M. G., 14th edition, 4th, impression, Griffin, 1965, page xvi:
    "The earliest occurrence of the word statistics yet noted is in *The Elements of Universal Erudition*, by Baron J. F. Von Bielfeld, translated by W. Hooper, M.D. (3 Vols, London 1770)".
    (ii) According to the same source:
    "Statist is of much earlier date than either Statistic or Statistics; the word statist is found, for example in William Shakespeare's play *Hamlet* (1602).

    (b) A Statist is defined in the Concise Oxford English Dictionary, 9th edition, 1995, as either:
    (1) A statistician, or
    (2) A supporter of statism, which in turn is defined as 'centralized State administration and control of social and economic affairs'.

Worksheet 2 (Solutions)

1. (a) -1.8
    (b) 3
    (c) 33.58
    (d) 0.000335
    (e) 0.1575
    (f) 341.297
    (g) 8.35
    (h) 5.29
    (i) 0.125
    (j) 0.0164
    (k) 1
    (l) 125
    (m) 4.953
    (n) 0.2019
    (o) 0.839

(p) 8

(q) 24, 1, 720, 1, not defined, not defined.

2. (a) 33.6

    (b) 0.00033

    (c) 0.16

    (d) 341.3

    (e) 300

3. (a) 55

    (b) 3

    (c) 55

    (d) 44

4. 24, 3, 576, 84, 0, 12, 12

5. 17.7, 3.54, 313.29, 85.27, 0, 22.612, 22.612

6. For the 8 values in Question 4:

$$\Sigma x = 24, \qquad \bar{x} = \frac{24}{8} = 3.$$

$$\Sigma(x - \bar{x}) = (2 - 3) + (3 - 3) + (5 - 3) + (1 - 3) + (4 - 3)$$
$$+ (3 - 3) + (2 - 3) + (4 - 3) = 0.$$

$$\Sigma(x - \bar{x})^2 = (-1)^2 + 0^2 + \dots + 1^2 = 12.$$

$$\Sigma x^2 - \frac{(\Sigma x)^2}{n} = (2^2 + 3^2 + \dots + 4^2) - \frac{24^2}{8} = 84 - 72 = 12.$$

This shows that the two results hold for the data in Question 4. Similarly, for Question 5 data,

$$\Sigma x = 17.7, \qquad \bar{x} = \frac{17.7}{5} = 3.54$$

$$\Sigma(x - \bar{x}) = (2.3 - 3.54) + \dots + (2.3 - 3.54) = 0.$$

$$\Sigma x^2 = 85.27 \quad \frac{(\Sigma x)^2}{5} = \frac{17.7^2}{5} = 65.658 \quad \text{and} \quad 85.27 - 62.658 = 22.61$$

Also, $\Sigma(x - \bar{x})^2 = (2.3 - 3.54)^2 + \dots + (2.3 - 3.54)^2 = 22.61$

Proofs:

$\Sigma(x - \bar{x}) = \Sigma x - \Sigma\bar{x} = (n\bar{x} - n\bar{x}) = 0.$ This uses the result that since

$$\bar{x} = \frac{\Sigma x}{n} \qquad \text{then} \qquad \Sigma x = n\bar{x}$$

$$\Sigma(x - \bar{x})^2 = \Sigma(x^2 - 2x\bar{x} + \bar{x}^2) = \Sigma(x^2 - 2\bar{x} \ \Sigma x + n\bar{x}^2)$$

$$= \Sigma x^2 - 2\frac{\Sigma x}{n} \ x + n\frac{(\Sigma x)^2}{n^2}$$

$$= \Sigma x^2 - 2\frac{(\Sigma x)^2}{n} + \frac{(\Sigma x)^2}{n} = \Sigma x^2 - \frac{(\Sigma x)^2}{n}$$

Worksheet 3   (Solutions)

1. (a) Grouped frequency table and bar chart.
    (b) List of countries in order of 'Number of earthquakes per unit area', and bar chart.
    (c) Grouped frequency table and histogram.*
    (d) Grouped frequency table and histogram.*
    (e) Grouped frequency table with 11 groups and line chart.
    (f) Grouped frequency table and histogram.*
    (g) Grouped frequency table with 6 groups and line chart.
    (h) Grouped frequency table with 6 groups and bar chart.
    (i) Simple list and dotplot.
    (j) Grouped frequency table with 3 groups and bar chart.
    (k) Grouped frequency table and histogram.*
2. The dotplot shows that the weight of coffee above the nominal 200 g varies from 0.2 to 5.7 g. The distribution of the weight is fairly symmetrical, except for two 'outliers' at 4.6 and 5.7, and is bunched in the middle, with a mean of about 2 g.
    Grouped Frequancy Table:

Weight (g)	Number of jars
0.00–0.99	13
1.00–1.99	25
2.00–2.99	20
3.00–3.99	9
4.00–4.99	2
5.00–5.99	1

The histogram drawn from the table above shows some positive skewness, see Fig. B.1.

---

* Instead of histogram, either a stem and leaf display or a box and whisker plot could be used.

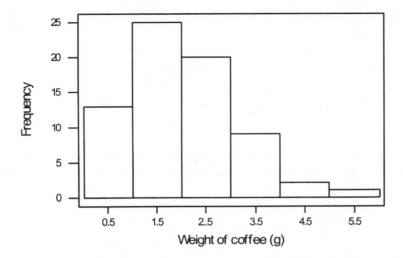

Figure B.1   Histogram of Excess Weight in 70 Jars of Coffee

However, if you now use Minitab to draw a histogram by:

Choose **Graph** > **Character Graph** > **Histogram**, the result is unsatisfactory because, not only does Minitab choose 12 groups instead of the 6 you see above, but the first group it chooses is from −2.5 to +2.5 with a mid-point of 0.0. This is not a good choice of the groups, since there should never be a negative value for the surplus of coffee above 200 g, since this is illegal!

Instead you should enter 0.5:5.5/1.0 in the box called **'define intervals using values'**.

The stem and leaf display is as follows:

```
    3        0 234
   13        0 5667778999
   24        1 11223333344
  (14)       1 56666677778889
   32        2 00112222333
   21        2 577788899
   12        3 011234
    6        3 557
    3        4 0
    2        4 6
    1        5
    1        5 7
```

Figure B.2   Stem and Leaf Display of Weights of Coffee Above 200 g in 70 Jars

The box and whisker plot is as follows:

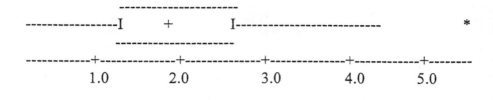

Figure B.3 Box and Whisker Plot for Weights of Coffee Above 200 g in 70 Jars

3.

Table B.1   Grouped Frequency Distribution for the Distance from Home to Oxford of 40 Students

Distance (km)	Number of Students
0–49.9	8
50–99.9	12
100–149.9	4
150–199.9	4
200–249.9	4
250–299.9	4
300–349.9	0
350–399.9	0
400–449.9	2
450–499.9	2

The question as to which of the 3 graphs you prefer is subjective. Here is my answer:

At this stage I rule out the box and whisker plot (see Fig. B.3) because it involves numerical measures we have not discussed yet (but we will in Chapter 4). The stem and leaf display, Fig. B.2, has the advantage that we still have all 70 of the raw data values, and we can see the shape of the distribution of weight by turning Fig. B.2 through 90 degrees, but unless you wanted to find the median weight, also to be discussed in the next chapter, I cannot see why you want so much detail in 'a summary'.

On the other hand, perhaps Fig. B.1, the histogram, has lost too much detail since there are only 6 groups. So, I conclude that the best graph would be a histogram with the first group being from 0.00 to 0.49 and so on. Try it!

Table B.2 Cumulative Frequency Distribution Table Using the Data in Table B.1

Distance (km)		Cumulative Number of Students
Less than	0	0
	50	8
	100	20
	150	24
	200	28
	250	32
	300	36
	350	36
	400	36
	450	38
	500	40

Half of the 40 students, i.e., 20, live less than 100 km from Oxford, according to Table B.2. From the raw data, the 20th value, when the 40 values are placed in order, is 96. The 21st value 104, so again the middle value is 100 km. This type of 'middle value' is called the 'median', and will be discussed in Chapter 4.

4. There are 31 BSc and 9 BA students. Using dotplots, we can say that the A-level count is generally lower for BSc students. (Approx. means are 11 and 16, although one BA outlier means that we should be comparing medians instead. Approx. medians are 10 and 13, respectively.) For a more sophisticated analysis of these data, see Sections 9.11 and 9.13 in this book.

5. (a) For all 86 clubs:

Table B.3 Number of Goals Scored on One Day by 86 Teams

Number of Goals	Number of Teams
0	18
1	35
2	22
3	8
4	2
5	1
Total	86

(b) Similar tables may be drawn for each of the four leagues All five graphs are positively skew, partly because you cannot have a negative number of goals!

(c) The mean number of goals scored in each of the four divisions is as follows:

	Mean Goals per Team	Mean Goals per Match
Prem	1.00	2.00
Div 1	1.50	3.00
Div 2	1.33	2.67
Div 3	1.44	2.88

The means are fairly similar, with a possible trend for higher scores on average as you go down the divisions. However, it isn't a perfect trend. I would say that the most likely effect is for fewer goals on average in the Premiership compared with the 3 'Nationwide divisions'.

(d) The 43 estimates of 'home advantage' are

Premier	Div 1	Div 2	Div 3
1	3	2	2
0	1	0	3
2	3	1	−2
0	0	1	0
2	2	−1	0
1	3	−1	0
0	0	1	−1
0	0	0	2
	1	2	0
	1	0	1
	1	−2	0
		1	2

Forming a frequency distribution:

Advantage (x)	Number of Matches (f)
−2	2
−1	3
0	15
1	11
2	8
3	4

The mean advantage $= \dfrac{\text{Total advantage}}{\text{Number of matches}}$

$$= \frac{(2x(-2) + (-1) + 11x1 + 8x2 + 4x3)}{43}$$

$$= \frac{32}{43}$$

$$= 0.74 \text{ goals}$$

This implies that playing at home is worth 3 goals in every 4 matches, but common sense tells us that we need more data, in case this was an unusual day for some reason, e.g., weather, an unusual number of 'local Derby's', or a lot of games in which one team outclassed their opponents. Even if we decided that this was a representative sample of matches, we would probably want to use the idea of a confidence interval for the mean advantage (see Chapter 9 in due course).

Worksheet 4   (Solutions)

1. (b)
2. (a)
3. (a)
4. (b)
5. (a)
6. (b)
7. (c)
8. (a) They measure a middle value for a set of (one-variable) data.
    (b) Sample mode.
    (c) Sample median.
    (d) Markedly skew data.
    (e) Only for categorical data (for which the mean is not defined).
    (f) For roughly symmetrical data.

9. (a) Because averages can be misleading, and to show how much a set of data vary.
   (b) (i) Sample standard deviation; (ii) Sample inter-quartile range.
   (c) The 12 weekly incomes in Question 10 (below).
   (d) (i) Sample standard deviation; (ii) Sample inter-quartile range; (iii) There isn't one!
   (e) (i) Sample mean; (ii) Sample median; (iii) There isn't one!

10. (a) Sample mean = 78.1, sample median = 67, sample mode = 62 or 67. These answers differ because of one extremely high value, which implies positive skewness, so we expect mean > median. The mode is non-unique and useless in this case.
    (b)

$$ s = \sqrt{\dfrac{89797 - \dfrac{937^2}{12}}{11}} = 38.9 $$

Sample inter-quartile range = upper quartile − lower quartile
$$ = 73.25 - 62 $$
$$ = 11.25 $$

These answers differ because the standard deviation and the I.Q.R. measure variation in different ways.

    (c) Coefficient of skewness = $3(78.1 - 67)/38.9 = 0.856$. This is fairly close to the guideline value of 1, which would imply markedly skew data. Hence it may be better to quote the median and I.Q.R. Alternatively, you could say that of 12 weekly incomes 1 was 200 and the other 11 had a mean of 67 and standard deviation of 6.47. (Notice the large effect on s.d. when the 'outlier' is left out of the calculation.)

11.

$$ \bar{x} = \frac{11.44}{11} = 1.04 \quad \text{and} \quad s = \sqrt{\dfrac{11.9068 - \dfrac{11.44^2}{11}}{10}} = 0.0303 $$

N.B. Do not round down 11.44 to 11.4 in the calculation of *s*. If you are curious, see what happens if you do!

12. It would have been a good idea to put all the 'coffee data' from Question 2 of Worksheet 3 into a computer file called, say, 'coffee', as explained in Appendix E. Then you should find that:
    (a) Sample mean = 1.98, and Sample standard deviation = 1.07.
    (b) Sample median = 1.8, and Sample inter-quartile range = 1.45 Measure of skewness = $3(1.98 - 1.8)/1.07 = 0.50$, which means that there is some positive skewness, but it is not marked.

In this example we would choose the mean as the preferred average and the standard deviation as the preferred measure of variation.

N.B. If you are thinking of using a calculator only in this question, you will have to type 70 values into your calculator to find the mean and standard deviation, while the best way to obtain the median and I.Q.R. is from the stem and leaf display (see solution to Question 2 of Worksheet 3).

13. For this question the solution will be made much easier by the use of Minitab. From Section 3.2, we know that the Distance data from Table 1.1 is in the computer file ES4DAT.MTW. We can go into Minitab and retrieve it. Then :

Choose **Stat** > **Basic Statistics** > **Descriptive Statistics**,
and then specifying **Distance** will produce all the summary statistics for Question 13:

Mean = 147.5, standard deviation = 129.4, median = 93.0,
Inter-quartile range = Q3 − Q1 = 220 − 50 = 170.
Measure of skewness = 3(147.5 − 93.0)/129.4 = 1.26

Since 1.26 > 1, we conclude that the distribution is markedly positively skewed, and we prefer the median and I.Q.R. as our summary statistics.

14. In this Question, we will see how to compare two groups of subjects for one particular variable. We will compare 'distances' for male and female students:

Choose **Stats** > **Basic Statistics** > **Descriptive Statistics**.
Then, as well as entering **Distance**, we click on the little box next to the word 'by' and enter **C2** in the larger box next to 'by'.

This gives Minitab output as follows:

SEX	N	MEAN	MEDIAN	STDEV	Q1	Q3
Males	13	167.1	141.0	131.0	62.5	249.5
Females	27	138.1	90.0	130.1	50.0	208.0

The coefficients of skewness for male and female distances are
Males: 3(167.1 − 141.0)/131.0 = 0.60,
Females: 3(138.1 − 90)/130.1 = 1.11.

Clearly there is more skewness in the female data, and it would have been better to use the median and I.Q.R. as summary statistics. The interesting question now is the difference between the median distances from home between the two groups due to a 'real' effect (females wishing to stay close to their 'roots'), or is the difference due to chance, and the small sample sizes involved. We will be in a much stronger position to answer these and other interesting questions when we have studied Chapters 9, 10, and 11.

15. We refer back to Table B.3 in this Appendix, which we now reproduce for convenience here:

Number of Goals	Number of Teams
0	18
1	35
2	22
3	8
4	2
5	1

We require a suitable 'average'. We rule out the mode for reasons that should now be obvious. The median is halfway between the 43rd and the 44th when the goals are arranged in order … 18 zeros, 35 ones, and so on. So the median = 1. The mean = $(18 \times 0 + 15 \times 1 + \cdots 1 \times 5)/86 = 116/86 = 1.35$. There is some skewness but is it marked? We need the standard deviation! If you type the 86 numbers into a calculator, you should find that $s = 1.06$. Using Formula (4.8), a measure of skewness $= 3(1.35 - 1)/1.06 = 0.99$. This creates a problem!, because this book gives the value 1 as representing 'marked skewness'. But this was only a guideline, a 'rough' guide. Personally, I prefer the mean in this situation because the median is sure to be an integer, i.e., a whole number like 1 or 2 or 3, so it jumps up in large steps, while the mean does not except when the odd outlier occurs.

Worksheet 5 (Solutions)

1. (c)
2. (b)
3. (b)
4. (a)
5. (a)
6. (c)
7. (b)
8. (c)
9. (b)
10. (a)
11. Refer to Sections 5.3 and 5.4.

12. Using $P(E') = 1 - P(E)$, where E and E' are success and failure, respectively, P(failure) $= 1 - $ P(success) $= 1 - 0.2 = 0.8$.

13. The three events '2 heads', '1 head and 1 tail' and '2 tails' are not equally likely. However, the events HH, HT, TH, and TT are equally likely and each has a probability of 1/4 of occurring:
    P(2 heads) = P(HH) = 1/4
    P(1 head and 1 tail) = P(HT or TH)
                         = P(HT) + P(TH), since events HT and
                           TH are mutually exclusive.
                         = 1/4 + 1/4
                         = 1/2
    P(2 tails) = P(TT) = 1/4.

14. Our estimate of the probability of heads is 1 after 5 tosses, but the 'number of trials' is not large enough for an accurate estimate to be made.

15. (a) Each die has the same probability (= 1/3) of being selected.
    (b) Y1, Y2, …, Y6, B1, B2, …, B6, G1, G2, …, G6; 18 outcomes in all.
    (c) Yes.
    (d) 1/18.
    (e) (i) 6/18, (ii) 3/18, (iii) 9/18, (iv) 3/18, (v) 12/18.

16. (a) 9/27, (b) 1, where 9/27 is the area of rectangle with base 164.5 to 169.5/Total area of histogram.

17.

Block Number	P(win)
1	12/50 = 0.24
2	33/100 = 0.30
3	50/150 = 0.33
4	65/200 = 0.325
5	86/250 = 0.344
6	107/300 = 0.357
7	122/350 = 0.349
8	140/400 = 0.350
9	155/450 = 0.344
10	172/500 = 0.344

(a) It depends how accurate you wish your estimate of the probability to be. It seems to be settling down after 250–350 games. In Section 9.9, we will look at the question of the sample size needed to estimate a binomial probability. (For example, with 500 games played we can estimate the true probability of

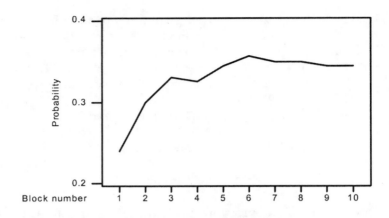

Figure B.4   Probability of Winning at Patience

winning to within ±0.04, for 95% confidence. How we do this and what 95% confidence means will have to wait for the moment.

(b) The games should be independent, as long as the cards are well shuffled after each game, and assuming that there is no 'learning process', or 'boredom factor'.

18. (a) A', (b) $A|B$, (c) $B|A$

19. (a) Probability that event A will occur, given that event B has already occurred.

(b) Probability that event B will occur, given that event A has already occurred.

(c) Probability that event A will not occur.

(d) Events A and B are independent if, when either occurs, the probability that the other will occur is not affected.

(e) Events A and B are mutually exclusive if, when either of them occurs, the other cannot.

20. $P(A \text{ and } B) = P(A)P(B|A)$

If A and B are statistically independent, $P(A \text{ and } B) = P(A)P(B)$

21. $P(A \text{ or } B \text{ or both}) = P(A) + P(B) - P(A \text{ and } B)$.

If A and B are mutually exclusive, $P(A \text{ or } B) = P(A) + P(B)$.

22. (a) A and B are statistically independent.

(b) A and B are mutually exclusive.

23. $P(3 \text{ or } 6) = P(3) + P(6) = 1/6 + 1/6 = 1/3$.

24. $P(\text{red or picture or both}) = P(\text{red}) + P(\text{picture}) - P(\text{red and picture})$

$= 26/52 + 16/52 - 8/52$

$= 34/52$.

25. The probability tree has 8 branch-ends, each having a probability of 1/8; (a) 1/8, (b) 3/8, (c) 3/8, (d) 1/8.

26. P(at least one hit) = 1 − P(all miss) = $1 - \frac{1}{2} \times \frac{2}{3} \times \frac{3}{4} = \frac{3}{4}$.

27. P(exactly one defective)
    = $P(DD'D'D'$ or $D'DD'D$ or $D'D'DD'$ or $D'D'D'D)$.
    = $P(DD'D'D') + P(D'DD'D') + P(D'D'DD') + P(D'D'D'D)$
    = 0.03 × 0.97 × 0.97 × 0.97 plus 3 similar terms
    = 0.1095.

28. From the information given, using C = car, H = house, and the laws of probability, P(H) = 0.4, P(C) = 0.7, P(H and C) = 0.3.
    (a) P(H or C or both) = P(H) + P(C) − P(H and C)
    = 0.4 + 0.7 − 0.3
    = 0.8.

    (b) Since

$$P(H \text{ and } C) = P((C) \times P(H|C))$$
$$0.3 = 0.7 \times P(H|C)$$
$$P(H|C) = \frac{0.3}{0.7} = 0.43.$$

29. (a) 11/20 = 0.55, (b) 2/11 = 0.18.

Table B.4   Number and Type of Bedroom, With or Without Bathroom, in a Hotel

| | Bedroom type | | |
	S	D	Total
Bath	2	9	11
No bath	4	5	9
Total	6	14	20

30. (a) P(success) = P(lift-off success) × P(separation success given lift-off success) × P(mission completed given that separation success)
    = (1 − 0.1) (1 − 0.05) (1 − 0.03)
    = 0.8294.
    P(failure) = 1 − 0.8294 = 0.1706.

31. (a) 13/40
    (b) 27/40
    (c) 26/40
    (d) (i) 13/13, (ii) 12/27.
    (e) (i) 13/25, (ii) 0/15, no, since P(male) = 13/40, while P(male height > 165), and these are not equal.
    (f) 11/40
    (g) P(male or BSc or both)
    $$= P(male) + P(BSc) - P(male\ and\ BSc)$$
    $$= 13/40 + 31/40 - 11/40$$
    $$= 33/40.$$

32. P(at least one 6 with 4 rolls of a die)
    $$= 1 - P(\text{ no 6s with 4 rolls of a die})$$
    $$= 1 - \left(\tfrac{5}{6}\right)^4 = 1 - 0.4823 = 0.5177.$$
    P(at least one double 6 when 2 dice are rolled 24 times)
    $$= 1 - P(\text{no double 6 when 2 dice are rolled 24 times})$$
    $$= 1 - \left(\tfrac{35}{36}\right)^{24}$$
    $$= 1 - 0.5086$$
    $$= 0.4914.$$
    Neither of the answers is equal to $2/3 = 0.6667$.

33. I am going to use the convention on this question that P(S = 4) is simply written P(4), and so on.
    (a) Look at Fig. 5.11(b). From this we can state P(7) = 6/36 and P(11) = 2/36. So, P(7 or 11) = 8/36 = 2/9.
    (b) P(4 before 7) = P(4 on first toss) + (not 4 and not 7 on 1st) × P(4 on 2nd toss), and so on. P(not 4 and not 7) = 1 − 3/36 − 6/36 = 3/4.
    So, P(4 before 7) = 3/36(1 + 3/4 + (3/4)^2 + (3/4)^3 + ...)
    $$= 3/36(1/(1 - 3/4)),\ \text{using the series}$$
    $$\text{given for } x = 3/4$$
    $$= 1/3.$$
    (c) Similarly, P(5 before 7) = 2/5
    (d) P(6 before 7) = 5/11.
    By symmetry, P(10 before 7) = P(4 before 7) = 1/3,
    P(9 before 7) = P (5 before 7) = 2/5.
    P(8 before 7) = P(6 before 7) = 5/11.
    Hence P(Winning) = P(7 or 11 on 1st toss) + P(4 on 1st) P(4 before 7), and similar terms with 5, 6, 8, 9, 10, instead of 4.
    So, P(Winning) = 2/9 + 2(3/36 × 1/3 + 4/36 × 2/5 + 5/36 × 5/11)
    $$= 0.493.$$

Worksheet 6 (Solutions)

1. A Bernoulli trial is an experiment with only two possible outcomes, often referred to as 'success' and 'failure'. The probability of success is constant.
2. The parameter p stands for 'the probability of success' in each trial.
3. The choice is arbitrary.
4. The number of successes in n Bernoulli trials.
5. By checking the four conditions (Section 6.3).
6. Look at the answer to Question 4, and then let $n = 1$. So, a Bernoulli distribution can be thought of as a binomial with only one trial.
7. (a) 2; (b) 3; (c) 4; (d) 1.5, 0.866; (e) Yes, because p = 0.5.
8. The correct answer is (b), as explained below:
   Noting that Table C.1 gives probabilities of 'r or fewer successes', we also note that the complement of '5 or more successes' is '4 or fewer successes'. Using Table C.1 for $n = 10$, p = 0.5, and $r = 4$ gives 0.3770. So, P(5 or more successes) = 1 − 0.3770 = 0.6230.
9. The correct answer is (c), since there can be 0, 1, 2, or 3 successes in 3 trials.
10. Again from Table C.1, for $n = 20$, p = 0.25, and $r = 3$,
    P(3 or fewer successes) = 0.2252, so the answer is option (a).
11. 0.0625, 0.2500, 0.375, 0.2500, and 0.0625. These were obtained using a B(4, 0.5) distribution. The probabilities sum to 1 because the 5 events are mutually exclusive and exhaustive (refer to Section 5.11). The expected values are obtained by multiplying the probabilities by the total sample size, i.e., 200 in this case. So, expect 12 or 13 families to have no boys (since $0.0625 \times 200 = 12.5$), and so on.
12. The number of correct answers has a B(20, 0.2) distribution.
    (a) Mean = $np = 4$.
    (b) P(8 or more correct) = 1 − P(7 or fewer correct)
    $$= 1 - 0.9679$$
    $$= 0.0321.$$
    (c) P(at least one correct) = 1 − P(none correct out of 20)
    $$= 1 - 0.0115, \text{ using Table C.1}$$
    $$= 0.9885$$
13. (a) 0.9231; (b) 0.2611; (c) 0.0002.
14. 0.0313; 0.1563; 0.3125; 0.3125; 0.1563; 0.0313.
15. 0.3828, probability.
16. (a) Because the four conditions have been satisfied
    (b) n = 20, p = 0.2
    (c) (i) 0.0115; (ii) 0.9885

17. Using B(6, 0.1), (a) 0.5314, (b) 0.0486, where a trial is examining an egg, and a success is finding a cracked egg. For the last part, use the idea that a trial is now examining a box of six eggs, and a success is finding a box containing no cracked eggs. So, we must use a B(5, 0.5314) distribution.

    P(3 or more) = P(3) + P(4) + P(5) = 0.5587.

18. m is the mean number of random events per unit time or space. e is a constant, the number 2.718 …

    x is the variable number of random events per unit time or space, which can take values 0, 1, 2, …

19. $m = 4$ and hence variance = 4, and hence standard deviation = 2. For any Poisson distribution, the mean and the variance are equal.

20. In each unit of time (or space) the probability that the event will occur is the same.

21. P(2) = P(2 or fewer) − P(1 or fewer)

    = 0.6767 − 0.4060, using Table C.2 for $m = 2$ and $r = 2$ and 1,

    = 0.2707. So (a) is the correct answer.

22. P(at least 23 events) = 1 − P(22 or fewer)

    = 1 − 0.9997, for $m = 10$ and $r = 22$.

    = 0.0003. So (a) is the correct answer.

23. $m = 1/10 = 0.1$ breakdowns per week. 0.9048, 0.0905, 0.0045.

24. $m = 1/5 = 0.2$ misprints per page. P(0) = 0.8187, i.e., 81.87%. The expected number of pages having no misprints is 0.8187 × 500 = 409.

25. $m = 2.5$. (a) 8.21%, (b) 45.62%, (c) P(4 or more) = 0.2424 or 24.24%.

26. $m = 3$. (a) 0.0498, (b) 0.2240, (c) 0.5768.

27. 0.634, 0.6321. Good agreement, the first answer is only an approximation, but a very good one. The second answer is correct.

28. $m = np = 4$, P(0) = 0.0183.

29. (a) 0.0245, 0.318, not such good agreement (as in Question 27 above). Answers using Poisson approximation are less accurate for small probabilities.

    (b) 0.1849, 0.1755, good agreement.

30. The p.d.f. of the geometric distribution is:

    $$P(x) = (1 - p)^{x-1} + p, \quad \text{for} \quad x = 1, 2, 3, \ldots$$

    The sum of the probabilities = $P(1) + P(2) + P(3) + \cdots$

    $$= p + (1 - p)p + (1 - p)^2 p + \cdots$$

    $$= p(1 + (1 - p) + (1 - p)^2 \cdots)$$

    $$= \frac{p}{1 - (1 - p)}$$

    $$= 1.$$

Worksheet 7 (Solutions)

1. $z = \dfrac{x - \mu}{\sigma} = \dfrac{13 - 10}{4} = 0.75.$ $P(z \leq 0.75) = 0.7734$, from Table C.3.

2. Strictly speaking the answer is 0, since $z = 0$ when $x = \mu$. However, we are told that the variable is continuous, so we use the idea that $P(10) = P(9.5 \leq x \leq 10.5)$. For $x = 10.5$, $z = \dfrac{10.5 - 10}{2} = 0.25$. Using Table C.3, the area to the left of 10.5 is 0.5987. So the area to the right of $10.5 = 1 - 0.5987 = 0.4013$. By symmetry, the area to the left of 9.5 is also 0.4013. The required area is $1 - 0.4013 - 0.4013 = 0.1974$ to 4 dps, or 0.20 to 2 dps. So (c) is correct.

3. The correct answer is (a):
When $x = \mu + 1.645\sigma$, $z = \dfrac{\mu + 1.645\sigma - \mu}{\sigma} = 1.645$. Using tables, area to the left of $z = 1.645$ is 0.95, so area in the right-hand 'tail' is $1 - 0.95 = 0.05$. Then use symmetry.

4. (c).

5. The total area under the curve for the normal distribution is 1, for the rectangular distribution the total area of the rectangle is also 1.

6. (a) 4.75%; (b) 0.05%; (c) 95.2%. Expect 2, 0, and 48, to the nearest whole orange. When mean is reduced to 65, new answers are: (a) 0.05%; (b) 4.75%; (c) 95.2%. Expect 0, 2, and 48.

7. For the seven grades, percentages are 0.62%, 6.06%, 24.17%, 38.30%, 24.17%, 6.06%, and 0.62%. The total price of 10000 oranges is equal to $62 \times 4 + 606 \times 5 + \cdots + 62 \times 10 = 70,000\ p$. The mean price is 7 p.

8. Percentage rejected = 18.15%. With new mean of 0.395, percentage Rejected = 13.36%. This is a minimum because 0.395 is exactly halfway between the rejection values of 0.38 and 0.41.

9. (a) 203, (b) 19, (c) 778.

10. For a left-handed area of $1 - 0.15 = 0.85$, $z = 1.04$, using Table 3(a) in reverse. It follows that $\dfrac{(x - \mu)}{\sigma} = \dfrac{(85 - 65)}{\sigma} = 1.04$.
So, $\sigma = \dfrac{(85 - 65)}{1.04} = 19.2$ cm.
When $x = 50$ cm, $z = -0.78$ which corresponds to an area of 0.7823 to the right of $z = -0.78$, i.e., greater than 50 cm. The required answer is $(100 - 78.23)\% = 21.77\%$ of years (approximately 1 in 5 on average).

11. The data implies a normal distribution with a standard deviation of 12 and an area of 0.2 to the left of 90. Using Table C.3(a) for an area of 0.8, i.e., $(1 - 0.2)$, it follows that, at 90, $z = -0.84$. The minus sign is because 90 is below the mean. Hence $-0.84 = \dfrac{90 - \mu}{12}$.
So, $\mu = 90 + 12 \times 0.84 = 100.08$.
When $x = 125$, $z = \dfrac{125 - 100.08}{12} = 2.08$. From tables, area to the left of $z = 2.08$ is 0.9812. Hence only 2% earn more than £125 per week.

12. (a) 50%; (b) 95.25%; (c) 99.95%.
13. $z = -2.33$, $\frac{(x - \mu)}{\sigma} = \frac{(x - 172)}{8} = -2.33$, so $x = 172 - 8 \times (2.33) =$ 153.4
14. (a) 50%, (b) 95.45%. New target mean is 25.82 kg, 0.1% exceed 27.37 kg ($z = 3.1$).
15. Starting with $z = \frac{(x - \mu)}{\sigma}$ let $x = \mu + \sigma$. Then $z = 1$, no matter what values the mean and standard deviation take, and Table C.3(a) gives 0.8413 as the area to the left of $z = 1$. So the area in the right-hand 'tail' is $1 - 0.8413 = 0.1587$. By symmetry, the area to the left of $\mu - \sigma$ is also 0.1587, so the area between $\mu - \sigma$ and $\mu + \sigma$ is $1 - 0.1587 - 0.1587 = 0.6826$ or 68% approximately. The other two statements can be confirmed in the same way, since they involve $z = \pm2$ and $z = \pm3$.
16. Use the normal approximation to the binomial with $n = 60$, $p = 0.8$, so $\mu = 48$ and $\sigma = 3.1$.
    (a) $P(>49.5) = 1 - 0.6844 = 0.3136$.
    (b) $P(49.5 - 50.5) = 0.1066$.
    (c) $P(<49.5) = 0.6844$.
17. With a rectangular distribution, area to the right of $15 = 1/6$, so he is late 1 day in 6 on average. With a normal distribution for which $\mu = 25$ and $\sigma = 7.5$, $z = 1.33$. Hence from Table C.3(a), the area to the left of 15 is 0.9082. So, the area to the right of 15 $= 1 - 0.9082 = 0.0918$, and he is late 1 day in 11 on average (see Fig. B.5).

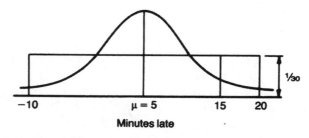

Figure B.5   Two Distributions of a Commuter's Lateness for Work

Worksheet 8   (Solutions)

1. (c)
2. (b)
3. (c)
4. (a)
5. (b)

6. (a), (b), (c), (d)—see Sections 8.1, 8.2, 8.3.
   (e) A census is a 100% sample, so the whole population is included in the sample, and often the main purpose is to count the total number of individuals in the population.
7. See Section 8.3.
8. (a) Might catch slowest and largest first; better to number the mice 1 to 20 and use random numbers.
   (b) (i) Travellers more affluent than the average adult
      (ii) Shoppers for food more representative of adults, might be biased in favour of housewives, etc.
      (iii) Adults leaving job centre may be unemployed, might be biased in connection with unemployment.
   (c) We do not know how the investigator actually 'randomly threw the quadrat'; better to use grid method (Fig. 8.1), and place the quadrat with its centre at the chosen points.
   (d) Use stratified sampling since there are three strata. Select 4, 5, 1 at random from the three types of hotel, respectively.
   (e) Initial sample correctly chosen, but many of those chosen may not have access to a telephone, in spite of the huge growth in the sale of 'mobiles'. Also telephones in houses may be used by more than one person, but only one name appears in the Phone Book. There is also the problem that some numbers are ex-directory.
   (f) People who visit a doctor's surgery may not be typical of all patients registered with him. Also, those who volunteer may do so because they are prone to influenza, again not random. It would be better to choose a random sample from the alphabetical list of patients, and assign half randomly to the vaccine and the other half to the placebo.
9. The following will simulate 108 throws of two dice, using Minitab:
   Type 1, 2, 3, 4, 5 and 6 into the first 6 rows of Column1 (C1). Type 0.1667 in each of the same rows of C_2.
   Choose **Calc > Random Data > Integer**
   Type **108** in **Generate** box.
   Type **C3 C4** in **Store in Columns** box
   Type **1** in **Minimum value** box, and **6** in **Maximum value** box
   Click on **OK**
   Choose **Calc > Mathematical Expressions**
   Type **C5** in **Variable** box
   Type **(C3 + C4)/2** in **Expression** box
   Click on **OK**

You should now have a random sample of 108 values in C3 and another in C4. Each of 108 rows in C5 should contain the mean, known as the 'score' in the question, of the two numbers in C3 and C4 in the same row. The mean and standard deviation of the numbers in C5 should be 3.5 and 1.21, respectively.
Check these as follows:

> Choose **Stat** > **Basic Statistics** > **Descriptive Statistics**
> Type **C5** in the **Variables** box
> Click on **OK**

Now obtain a histogram of the numbers in C5 as follows:

> Choose **Graph** > **Character Graph** > **Histogram**
> Type **C5** in **Variable** box

You might also wish to get a print-out of the session as follows:

> Choose **File** > **Print Window**

Repeating the simulation experiment for three dice means using C3, C4, and C5 for the initial simulation and C6 for the sample mean. The expected mean and s.d. for C6 are 3.5 and 0.98. The histogram of C6 should be more 'normal' than was the case for samples of size 2.

10. Before you come to the main list of questions on the questionnaire, ask the student to write down their name, student number and field(s) of study. Then ask the student to read a short sentence, already typed on the questionnaire, indicating why you are doing this project and also stating that the information they give will be treated confidentially.

(a) List of Questions

    (i)   Are you a full-time student? Yes or No?

    (ii) Which year are you in; 1st, 2nd, 3rd, or other (please specify)?

    (iii) Have you done more than five hours per week (on average) of paid employment while you have been a student at OBU?

    (iv) If yes, give details for the last three completed terms together with your average mark in each term. If there were special circumstances beyond your control for any mark to be non-typical, please give brief details.

    (v) If you said Yes to (iii) above, what were the two main reasons for working: (then make up a possible list of reasons, say 5 at most with a 6th stating 'other reason, please specify').

    (vi) If you said No to (iii) above, what were the two main reasons why you did not work. (Make up another list.)

(b) Sections 9.6 and 9.9 mention this topic, but it is tricky to discuss this in-depth at this stage. The pragmatic answer is to collect as much data as you can in the time available, which will include data analysis and writing a report.

(c) Students whose name begins with A, B, or C, say, may not be representative of the complete list of students; better to take 1 in every 10 from the list systematically, but choosing the first name randomly from the first 10 in the list.

(d) It may be too small. Some surveys like this have a response rate of only about 20%. Worse still, those who do reply may not be typical. You could send reminders, but it might have been better to stop students in the corridor on Campus, and ask the questions face to face. This method also enables follow-up questions to be asked to clarify the answers given verbally.

(e) Calculate the average mark for each student. You will then have two lists of marks, one for those who did work and a second for those who did not. Compare with histograms, one above the other just to get a feel for any differences. Also obtain the mean, median, s.d., and IQR for each group ('work' and 'did not work').

(f) The best analysis is probably an unpaired t test , but this will not be covered until Chapter 10. Your answers to parts (i), (ii), (iii) are subjective. Clearly, the bigger the difference in the means the more likely you are to conclude that it is a REAL effect, and not due to chance.

(g) Pilot Survey. Yes, do one if time. It may make the questions easier to answer, and the response rate is important. Also, you will know how long one interview takes if you decide the e-mail method has a poor response rate. Another problem is whether students will disclose their marks to 'a stranger'. Would you?

Worksheet 9   (Solutions)

1. To give a measure of precision to a single value (or point) estimate of a population parameter, such as the mean (μ) or the proportion of successes (binomial p).
2. (c).
3. False.
4. True.
5. (a) $\bar{x} \pm 2.58 \frac{s}{\sqrt{n}}$
   (b) $\bar{x} \pm 1.645 \frac{s}{\sqrt{n}}$
6. True.
7. It does NOT imply '95% probability', since either the population mean does lie between 10 and 12 and then the probability is 1,

or it does not and the probability is 0. Since, in repeated sampling, 95% of the 95% confidence intervals we calculate actually contain the mean, we feel that the confidence interval '10 to 12' has a very good chance of being one of those intervals which actually contains the population mean. Think of betting on a horse at odds of 19 to 1 on. Alternatively, we can think of taking a risk of 5% that the interval does not contain the population mean.

8. A 95% confidence interval for the mean of all accounts is

$$30 \pm \frac{1.96 \times 10}{\sqrt{100}}, \quad \text{or} \quad 30 \pm 2, \quad \text{i.e., 28 to 32.}$$

The required number of customers is n, where

$$\frac{1.96 \times 10}{\sqrt{n}} = 1. \text{ Hence } n = (1.96 \times 10)^2 = 384 \text{ customers.}$$

9. After 50 games of Patience in which there were 12 'wins', a 95% confidence interval for p is given by $\frac{12}{50} \pm 1.96 \sqrt{\frac{0.24 \times (0.76)}{50}}$, which is 0.24 ± 0.118, or 0.12 to 0.36 to 2 *dps*. For $n = 100, 200, 300$, and 500 we obtain 0.24 to 0.42; 0.26 to 0.38; 0.31 to 0.41; 0.30 to 0.38. Notice the width of the intervals getting smaller as n increases. If the error term is 0.03, the required number of games is n where:

$$0.03 = 1.96 \sqrt{\frac{0.344 \times 0.656}{n}},$$

from which we find $n = 963$.

10. 125.3 to 132.5; (a) 7.2, (b) 9.5, (c) 3.6.
11. 9220 pebbles.
12. $\bar{x} = 308.3, s = 131.9 \quad 274.2 \text{ to } 342.4.$
13. 0.67 to 6.75 kg. The assumption that difference in weights is approximately normal is reasonable since weight is approximately normal. Also a dotplot of the differences indicates symmetry and a concentration of points in the middle.
14. $s = 0.0982, -0.37$ to -0.58 for $(\mu_A - \mu_B)$. Assumptions: (i) percentages are normally distributed, (ii) $\sigma_A = \sigma_B$, reasonable here since s_A and s_B are similar (see also solutions to Questions 21, 22, and 23 of Worksheet 10).
15. $\bar{d} = -10.45, s_d = 10.13, -3.6$ to -17.3 for μ_d, where $d =$ rainfall in A $-$ rainfall in B.
The assumption that d is normal looks reasonable from a dotplot.

16. $s = 4.00$, -4.0 to -14.3 for $\mu_A - \mu_B$. Assumtions: (i) weights are normally distributed, a reasonable assumption if the reasons for small variations are numerous and independent (see Section 7.2). (vii) $\sigma_A = \sigma_B$. (Refer to solution to Question 14 above.)
    (a) 440.7 to 448.6.
    (b) 449.4 to 458.3.
    Reasonable for B, not for A, since 95% confidence interval for μ_B does contain 452, but 95% confidence interval for μ_A does not contain 452 (see also Section 10.17, where this use of confidence intervals is discussed).

17. There is NO solution to this Question, because it isn't really a question!

Worksheet 10 (Solutions)

1. See Sections 10.2 and 10.3.
2. 3. and 4. See Sections 10.4 and 10.15.
5. See Section 10.3.
6. If we wish to decide whether the value of the parameter of interest is greater than (or less than) a particular value, then the alternative hypothesis is one-sided. If we wish to decide whether the value of the parameter is different from (i.e., not equal to) a particular value, so the direction of the difference is not of interest, then the alternative hypothesis is two-sided.
7. (a)
8. (c)
9. (b)
10. (c)
11. (c)
12. *Calc t* $= 4.37$, a two-sided alternative hypothesis, *Tab t* $= 2.228$, $p = 0.0014$; data do not support stated hypothesis. Assumption: weight of sugar is approximately normal, a reasonable assumption if the reasons for small variations are numerous and independent (see Section 7.2). Alternatively, we could draw a dotplot (see Fig. 3.1) to see if the data are symmetrical and concentrated in the middle. The dotplot is, in fact, reasonably symmetrical with an even spread, so can be assumed to be 'approximately normal'.
13. H_0: $\mu = 0.30$, H_1: $\mu > 0.30$. If we reject the null hypothesis, this implies that we accept the alternative, which means that the manufacturer's claim is not justified. This is what happens in this case since *Calc t* $= 10.54$, *Tab t* $= 1.645$.

Assumption: nicotine content is normally distributed, but it is not important here since the sample size is so large. Note that Minitab cannot be used in this question since we are not given the 1000 individual values of nicotine (i.e., the 'raw' data).

14. *Calc* $t = -1.18$, $H_1: \mu < 110$, where μ is the population mean take-home pay for all farm workers. *Tab* $t = 1.699$, $p = 0.12$, data do not support claim. Assumption: approximate normality of wages, which a histogram would indicate. In any case the sample size is quite large.

15. The question implies that the market share is still 30% will be our null hypothesis, while the alternative is that it has increased. *Calc* $z = 1.09$, *Tab* $z = 1.645$. Market has not increased significantly. Assumption: the four binomial assumptions, the main one being independence, which implies that individuals do not influence each other in their choice of brand.

16. *Calc* $z = 3.46$, *Tab* $z = 1.96$. This is clearly a two-sided test since we are looking for a difference in the percentage in either direction from 50%. Since $|3.46| > 1.96$, $H_0: p = 0.5$ is rejected in favour of $H_1: p \neq 0.5$. It is not reasonable to expect that 50% of all gourmets prefer thin soup. Assumption: similar to Question 15 above.

17. *Calc* $z = -2.55$, *Tab* $z = 1.645$. It is reasonable to conclude that the death rate is lower than 14%, choosing a one-sided alternative. Assumption: the four binomial conditions, the main one being independence, which means that the risk of death is not influenced by others dying—probably true except for epidemics of fatal diseases (the Great Plague, for example).

18. (a) Growing conditions within a farm will be more homogeneous than between farms.
    (b) $\bar{d} = 0.3143$, $s_d = 0.4140$, *Calc* $t = 2.01$, *Tab* $t = 2.447$. $p = 0.091$.
    Mean yields are not significantly different.
    Assumption: approximately normal differences; reasonable if reasons for small variations in differences are numerous and independent (see Section 7.2). As in Question 12, a dotplot could be drawn, but would not be conclusive.

19. The last sentence of the question states 'less than', so H_1 is one-sided. $\bar{d} = -0.66$, $s_d = 0.8591$, *Calc* $t = -1.72$, *Tab* $t = 2.132$, $p = 0.16$. Allegation not supported by the data.
    Assumption: approximately normal differences. No information on why variations occur, and very little data, so t test dodgy here.

20. $\bar{d} = 0.72$, $s = 0.7941$, *Calc* $t = 2.87$, *Tab* $t = 1.833$, $p = 0.0093$. Drug gives significantly more hours of sleep. Similar dotplot to that for Question 12.

21. $\bar{x}_1 = 4673.3$, $s_1 = 120.94$, $\bar{x}_2 = 4370.0$, $s_2 = 214.48$, $s = 174.1$, *Calc t* = 3.02, *Tab t* = 2.228, $p = 0.013$.
    Here, H_0: $\mu_1 = \mu_2$, H_1: $\mu_1 \neq \mu_2$.
    There is a significant difference in the mean strengths.
    Assumptions: (a) approximate normality in both populations, difficult to tell here and very little data. (b) $\sigma_1 = \sigma_2$. *Calc F* = $\frac{s_2^2}{s_1^2}$ = 3.15, *Tab F* = 5.05 (Table C.6 for 5, d.f.). Since $3.15 < 5.05$, assumption of equal variances is reasonable (assuming normality, which is problematic as we have seen!).

22. $s = 1.1045$, *Calc t* = 3.777, *Tab t* = 2.01. It is reasonable to suppose that corner shops are charging more on average than supermarkets.
    Assumptions: (a) approximate normality, not important here because of large sample sizes (except as an assumption of the F test). (b) $\sigma_1 = \sigma_2$. *Calc F* = 1.44. *Tab F* = 1.98, for 24, 24 d.f. (Table C.6). Since $1.44 < 1.98$, assumption (a), of normality, is reasonable. We would need dotplots of the raw data to check visually for normality, but the sample sizes, 25 and 25, are reasonably large.

23. $\bar{x}_A = 94.00$, $s_A = 28.17$, $n_A = 10$, $\bar{x}_B = 99.00$, $s_B = 27.47$, $n_B = 10$, $s = 27.82$, *Calc t* = -0.40, *Tab t* = 1.73. Mean amount of vanadium for area A is not significantly less than for area B.
    Assumptions: (a) approximate normality in each population, difficult to tell here with small sample sizes, but dotplots do not indicate any extreme values and there is approximate symmetry. (b) $\sigma_A = \sigma_B$, *Calc F* = 1.05, *Tab F* = 3.19, so reasonable assumption.

24. (12) 1.02 to 1.06. Reject H_0: $\mu = 1$ since 1 is outside the 95% confidence interval.
    (16) 54% to 66%. Reject H_0: $p = 0.5$, Reject since 50% is outside the 95% confidence interval.
    (18) -0.07 to 0.70. Do not reject H_0: $\mu_d = 0$, since 0 is inside the 95% confidence interval.
    (21) 79.3 to 527.3. Reject H_0: $\mu_1 = \mu_2$, since 0 is outside the 95% confidence interval.

25. Ideally you would have two lists of average marks, one for those who did paid work and one for those who did not, rather like Questions 21 and 23 in this worksheet. So, you would want to draw two dotplots to look for approximate normality. Then carry out an F test to decide whether the assumption that the population variances are equal is justified. If both assumptions seem reasonable, carry out an unpaired *t* test. If there is a significant difference between the means, it may be because doing paid work has a detrimental effect on students' academic performance. Of course it may have an effect in the opposite direction which could be difficult to explain! What can we conclude if the difference in the

means is 'not significant at the 5% level'? One possible answer is that there really is no difference! Another is that the sample sizes were too small, but it was argued that your sample sizes should be as large as possible (in the solution to Question 10 of Worksheet 8). A third possibility is that other variables affect the effect of paid work on average marks. No two students are identical and there is sure to be variation between students even if they (i) either all worked or (ii) none of them worked. What can we conclude if the difference is significant and in the 'right' direction, i.e., the average for those who did paid work was lower than for the nonworkers? Even this is not clear-cut. For example, we might have had biased samples (as mentioned in Worksheet 8 Solutions). Also it is possible that the two groups differ in some previously unsuspected way; suppose those who do paid work contain a higher proportion of Arts students, who may have less contact time per week than Science students, since the latter may have a great deal of laboratory work to do. Finally, what if we suspect nonnormality and/or unequal variances? The answer to this is covered in the next chapter, Chapter 11!

Worksheet 11 (Solutions)

1. (c)
2. (a)
3. (c)
4. Hypotheses, assumptions.
5. Assumptions, powerful.
6. Null, alternative, higher.
7. Unpaired samples, powerful, assumptions, assumptions, standard-deviations (or variances).
8. (a) Preference testing example.
   (b) Examples where the magnitudes of the differences are known. Wilcoxon test preferred.
9. $p = 0.0039$, reject null hypothesis since $p < 0.05$, using Minitab. Using a calculator, you will be comparing 0.001953 with 0.025, with an identical conclusion.
10. The number of cigarettes with nicotine content greater than 0.30 mg.
11. Using Minitab, $p = 0.140$ is greater than 0.05, so null hypothesis is not rejected. If done by hand, *Calc T* = 3, *Tab T* = 0, same conclusion as Minitab's.
12. (a) Put data in C1 and C2 and put the differences in C3. Wilcoxon test, $p = 0.043$, null hypothesis rejected (close call, i.e., it is

only just significant at the 5% level). If Minitab is not used, *Calc T* = 1.5 while *Tab T* = 2, so null hypothesis is rejected.

(b) Mann-Whitney *Calc U* = 26.5, *Tab U* = 13. Difference is not significant. From Minitab, p = 0.5995 which is far greater than 0.05, hence do not reject null hypothesis. Put data into C1 and C2. The p value is 0.7913 according to Minitab, so null hypothesis is not rejected. By hand, *Calc U* = 46, *Tab U* = 37, so same conclusion as Minitab's.

13. Put data into C1 and C2. *Calc U* = 9, *Tab U* = 37, so there is a significant difference between brands. Minitab gives p = 0.0003, much smaller than 0.05, and hence significance confirmed.

14. The Mann-Whitney U test is appropriate here, using Minitab for Windows:

Enter the Brand1 data into the first 12 rows of column 1, (C1). Similarly for Brand2 and C2. Now, follow the example in Section 11.11. The Minitab output should include the following:

Test of ETA1 = ETA2 vs. ETA1? ETA2 is significant at 0.0003.

The test is significant at 0.0003 (adjusted for ties).
So we have a p value of 0.0003, which is much smaller than 0.05, and our conclusion could be that there is a significant difference between the 'mileages' of the two brands of tire. Looking again at the Minitab output, we see that the median mileages for the two brands are 44.5 and 37.5. So it is clear that Brand1 is significantly the better of the two brands at the 5% level (and also at the 1% and 0.1% level).

15. Since no data are provided, we can only discuss possibilities. As stated in the solution to Question 25 of Worksheet 10, ideally we would want to use an unpaired t test. However, if the assumptions of normality for both types of student, and/or the equality of the two variances, are in doubt, then it would be safer to use a Mann-Whitney U test.

Worksheet 12   (Solutions)

1. Analysis of Variance (ANOVA) table follows.
Since *Calc F* > *Tab F*, or since the p value is less than 0.05, the null hypothesis that the mean strengths of the two cements are equal should be rejected. We can therefore conclude that there is a significant difference between the mean strengths of the two cements.

Source of Variation	S.S.	d.f.	M.S.	Calc F	Tab F	p
Between types of cement	276033	1	276033	9.11	4.96	0.013
Within types of cement	303134	10	30313			
Total	579167	11				

This agrees with the solution to Question 21 of Worksheet 10 where we found *Calc t* = 3.02, *Tab t* = 2.228 and the same conclusion. Also notice that (a) the square of 3.02 is equal to 9.11 (the value of *Calc F*) and (b) the square of 2.228 is equal to 4.96 (the value of *Tab F*).

2. It is not possible to answer this question using Minitab because we do not have the raw data, i.e., the 50 individual prices. However, we can still apply Formulae 12.1 and 12.2 by using Formulae 2.1 and 4.4 to the summary statistics given. We will use suffix 1 for the corner shops and suffix 2 for supermarkets.
Since $\bar{x}_1 = \frac{\Sigma x_1}{n_1}$, $\Sigma x_1 = n_1 \bar{x}_1 = 486.25$. Similarly, $\Sigma x_2 = 456.75$.
Using Formula 4.4, squaring both sides gives

$$s_1^2 = \frac{\Sigma x_1^2 - \frac{(\Sigma x_1)^2}{n_1}}{n_1 - 1}$$

which can be rewritten as

$$\Sigma x_1^2 = (n_1 - 1)s_1^2 + \frac{(\Sigma x_1)^2}{n_1} = 9886.2$$

Similarly, $\Sigma x_2^2 = 8368.8$, and using Formulae 12.1 and 12.2, we can draw up the ANOVA table:

Source of Variation	S.S.	d.f.	M.S.	Calc F	Tab F
Between types of shop	17.405	1	17.405	14.28	4.04
Within types of shop	75.920	48	1.219		
Total	82.300	49			

Since $14.28 > 4.04$, we conclude that H_0 should be rejected, and that there is a significant difference between the mean prices in the two types of shop.

When we did a t test on the same data, we found that $Calc\ t = 3.777$ and $Tab\ t = 2.01$. Since $3.777 > 2.01$, the null hypothesis was rejected, which checks with the F test above. In fact we have even better checks since $3.777^2 = 14.27$ and $2.01^2 = 4.04$.

3. The ANOVA table is

Source of Variation	S.S.	d.f.	M.S.	Calc F	Tab F	p
Between Areas	125	1	125	0.161	4.43	0.692
Within Areas	13930	18	774			
Total	14055	19				

This time we do not reject the null hypothesis because $p > 0.05$, or because $Calc\ F < Tab\ F$. We conclude that there is no significance between the mean amounts of vanadium for the two areas A and B. In the t test done in Question 23 of Worksheet 10, we found that $Calc\ t = -0.40$ and $Tab\ t = 1.73$.

Now it is true that $(-0.40)^2 = 0.161$, at least approximately, but $(1.73)^2$ is not equal to 4.43. What has gone wrong? The answer is 'nothing'! In the t test we had a one-sided alternative hypothesis, while the corresponding F test above involves a two-sided alternative. We should have been comparing 4.43 with the square of 2.101 and not with the square of 1.73.

4. The ANOVA table is

Source of Variation	S.S.	d.f.	M.S.	Calc F	Tab F	p
Between Groups	3088.9	2	154.4	6.28	3.64	0.010
Within Groups	3932.8	16	245.8			
Total	7021.7	18				

Since $p < 0.05$, reject the null hypothesis (or say, 'since $Calc\ F > Tab\ F$ reject the null hypothesis').

The two assumptions are that:
  (i) Within each group the test scores are normally distributed.
  (ii) The (population) variance within each group is the same.
  On (i), three dotplots indicate that the data in each group

could be normal; at any rate it doesn't seem to be markedly nonnormal. On (ii), the standard deviations of the three samples are 15.6, 15.3, and 16.0. The fact that the sample variances are so close indicates that the population variances could well be equal.

Should we do a posterior test, since we rejected the null hypothesis? In theory, yes! But in practice let's see if the means of each group hint at possible conclusions from such a test. The means are

Group 1	Group 2	Group 3
99.5	106.8	129.0

The means of Groups 1 and 2 are much closer together than is the case for the other two pairs of Groups. The only sensible conclusion is that the mean score for Group 3 is significantly higher than for Groups 1 and 2.

Worksheet 13   (Solutions)

1. Numerical.
2. Contingency, individuals.
3. Independent.
4. Expected, $E = \dfrac{\text{row total} \times \text{column total}}{\text{grand total}}$.
5. 5, rejection.
6. $(r - 1)(c - 1)$, 1
7. 3.84
8. (b)
9. *Calc* $\chi^2 = 2.91$, *Tab* $\chi^2 = 3.84$. Since $2.91 < 3.84$, the null hypothesis of independence is not rejected. We conclude that there is no significant difference between the proportions of privately owned cars and company cars failing the M.O.T. Note that Minitab gives 4.00 instead of 2.91 for the calculated value of the test statistic. This is because Minitab does not use Yates's correction.
10. *Calc* $\chi^2 = 10.34$, *Tab* $\chi^2 = 7.82$. There is a significant difference between the proportion of male to female customers calling for petrol at the four garages.
11. Because of low E values, combine rows '30%–70%' and 'under 30%' to form a 2 × 2 table. *Calc* $\chi^2 = 1.52$, *Tab* $\chi^2 = 3.84$. Since $1.52 < 3.84$, we conclude that the chance of passing is independent of attendance (as defined by 'greater than 70%' or 'less than or equal to 70%').

12. *Calc* χ^2 = 5.48, *Tab* χ^2 = 5.99. The proportions of A, B, C are not significantly different (5% level).

13. Fisher exact test, since some E values are below 5 and we have a 2 × 2 table. So a = 1, b = 1, c = 9, d = 6. Probability = 0.3228, so do not reject the null hypothesis of independence (0.3228 > 0.05). The really keen student will note that the probabilities of the other five tables having the same marginal totals, but with a = 5, 3, 2, 1, 0, respectively, are 0.0830, 0.3874, 0.1761, 0.0293, 0.0014. This gives a total probability of 0.6126 for a two-sided alternative hypothesis, or 0.4508 for a one-sided alternative. In both cases, the null hypothesis is not rejected.

14. The standard χ^2 *test* gives *Calc* χ_2^2 = 18.31, while the Trend test gives *Calc* χ_1^2 = 14.82. Since 14.82 > 3.84, there is some evidence of a linear trend. Since *Calc* χ_3^2 = 18.31 − 14.82 = 3.49, which is less than 3.84, so we can conclude that there is no significant departure from linearity. The proportions across the three categories of patient are 64%, 64%, and 76%, respectively (had the middle value been 70%, this would have shown perfect linearity). The main conclusion is that, of those getting better, a higher proportion received drug A than was the case with the other two categories of patient. In short, A has a higher rate than B.

15. From the numbers given, the following 2 × 2 table can be drawn up:

	Enrolled	Did not Enroll
Interviewed	9	33
Not interviewed	1	25
Expected values are	6.2	35.8
	3.8	22.2

This means that, since 3.8 < 5, we should perform a Fisher exact test. Suppose we decide that 3.8 is 'near enough' to 5, so we could see what a χ^2 test would give. Using Yates' correction, it turns out that *Calc* χ^2 = 2.60. Since 2.60 < 3.84, the null hypothesis that interviewing and enrollment are independent is not rejected.

Now we will carry out a Fisher Exact test, because it is the correct test for this situation. For the table above containing the observed frequencies and using Formula 13.4, p_1 = 0.0399. We now form a second table by reducing the lowest number in the table above, i.e., 1, to 0, and calculate the other three entries in the table, while keeping the marginal values the same. So 9 becomes 10, 33 becomes 32, and 25 becomes 26. The resulting probability value is p_2 =

0.0051. The total probability is the sum of p_1 and p_2, which comes to $0.0399 + 0.0051 = 0.0452$. This is less than 0.05, so we now reject the null hypothesis (just!). Looking at the direction of the dependence, it is clear that $O > E$ for the cell for those who were interviewed and enrolled, since $9 > 6.2$. On the face of it, interviewing gained nearly 3 students.

Reservations: Just because we have found 'a significant dependence', we cannot conclude straightaway that there is cause and effect (we will see similar reservations when we tackle Correlation in Chapter 14). In this Question, it may be that those who chose to be interviewed were already predisposed to enroll in this university. The important practical question is "Shall we continue to interview in the future?" I would find it difficult to say "No", assuming that the Statistics Department 'needed the students'.

Worksheet 14   (Solutions)

1. Scatter diagram (or Scatter Plot, used in Minitab).
2. Correlation coefficient.
3. Correlation coefficient, Pearson, r, ρ.
4. r, arbitrary.
5. -1 to $+1$, negative, zero.
6. ρ, normally distributed, uncorrelated.
7. (c)
8. (a)
9. (b)
10. If x denotes the percentage increase in unemployment and y denotes the percentage increase in manufacturing output,
    $\Sigma x = 70$ $\Sigma x^2 = 1136$ $\Sigma y = -50$ $\Sigma y^2 = 664$ $\Sigma xy = -732$ $n = 10$
    $r = -0.739$, *Calc t* $= -3.10$, *Tab t* $= 1.86$ (or we could use *Calc r* $= -0.739$ and *Tab r* $= 0.6319$). Here our alternative hypothesis is two-sided, namely $H_1 : \rho \neq 0$. Since $|Calc\ t| > Tab\ t$, reject the null hypothesis that $\rho = 0$ (or say that $0.739 > 0.6319$, and reach the same conclusion). Hence, we can say that there is a significant correlation between x and y, which is clearly negative since the value of r is negative.
    Assumption: x and y are normally distributed. This assumption is seen as reasonable if a dotplot is drawn for x and another for y. It is possible to use the scatter diagram to judge approximate normality by first of all imagining that the points are projected on to the x axis, and then on to the y axis.
    It is, of course, difficult to draw conclusions about cause and effect. Does an increase in the increase in unemployment cause a decrease

in the increase in manufacturing output, or vice versa? Alternatively, are there other variables which affect both x and y?

11. The number of times that the commercial is shown is not normal. Use Spearman's r_s, which is 0.736 for these data (using the method of Section 14.7, because of ties). *Tab* r_s = 0.643, showing significant positive correlation. So the increase in the number of times the commercial is shown is associated with an increase in receipts. The scatter diagram gives the impression that the effect is flattening off after about 30 commercials in the month.

12. No evidence of nonnormality or normality, so safer to use Spearman's r_s, which is 0.468 (note ties) for these data. *Tab* r_s = 0.714. Larger areas are not significantly associated with longer river lengths.

13. No evidence of nonnormality or normality; r_s = −0.964 (no ties), *Tab* r_s = 0.714. Lower death rate is significantly associated with higher percentage using filtered water. Lower death rate could be due to other factors such as public awareness of the need to boil unfiltered water, better treatment of typhoid, for example.

14. Dotplots of income and savings indicate that while the distribution of income is reasonably normal, the distribution of savings may be positively skew. It is safer to use Spearman's r_s, which is −0.087 in this case. *Tab* r_s = 0.377, indicating no significant correlation between income and savings. There are two outliers in the scatter diagram, but the effect of leaving them out (which we should not do without a good reason) would probably make no difference to the conclusion above.

15. 'Height' and 'Distance from home to Oxford' are the two continuous variables. A dotplot of height is reasonably normal, but distance shows some positive skewness. In fact, the solution to Question 13 of Worksheet 4 gives a measure of skewness of 1.26, which indicates 'marked skewness' (see Section 4.13). It would be safer to use r_s in this case, and we will use the method of Section 14.7, since there is a tie in both variables. The data for the first ten students are set out below, in the order of increasing height. The ranks of height are set out in column 2, and the ranks of distance are in column 4. We will call the ranks of height x and the ranks of distance y. Spearman's r_s is found by calculating Pearson's r for x and y using Formula 14.1.

For our data,

$$r_s = \frac{303.25 - \frac{55 \times (55)}{10}}{\sqrt{\left[384.5 - \frac{55^2}{10}\right]\left[384.5 - \frac{55^2}{10}\right]}} = \frac{0.75}{82} = 0.0091$$

Height (ranked)	Ranks	Distance	Ranks
152	1	90	7
157	2.5	80	5.5
157	2.5	272	9
160	4	72	4
163	5	3	1
164	6	10	3
165	7	8	2
173	8	485	10
180	9	176	8
183	10	80	5.5

Now, *Tab* r_s = 0.648. Since 0.0091 < 0.648, the null hypothesis is not rejected. There is no significant association between student height and distance from home to Oxford.

16. (a) (i) The five points on the scatter diagram would lie on a straight line with a positive slope. An example are the points with (x, y) coordinates: (0, 5) (1, 7) (2, 9) (3, 11) (4, 13). These points lie on the line $y = 5 + 2x$, which has a slope of $+2$.

(ii) The points lie on a straight line with a negative slope. Let's use the line: $y = 5 - 2x$. Five points on this line are (0, 5) (1, 3) (2, 1) (3, −1) (4, −3).

(iii) For $r = 0$ exactly, we need a pattern of points which has no tendency for one of the variables to increase as the other increases, and no tendency to decrease as the other increases. Consider the points: (1, 2) (1, 8) (5, 2) (5, 8) and (3, 4). On the scatter diagram, these points appear like the four corners of a square together with the centre-point of the square. If Pearson's r is calculated, it will be equal to $\frac{0}{\sqrt{16 \times (107)}} = \frac{0}{41.38} = 0$.

(b) (i) The points (1, 3) (2, 3) (3, 3) (4, 3) and (5, 3) all lie on the line $y = 3$, which is parallel to the x axis. If Pearson's r is calculated, it will be equal to $\frac{0}{\sqrt{10 \times 0}} = \frac{0}{0} = ?$. This is a puzzle! The answer is that r in this case is 'indeterminate', which means it cannot be determined. This is not usually a problem in practice since such data are unlikely to occur. My intuition tells me that, if when x changes, y doesn't change at all, then the variables are independent and hence have zero correlation. What does yours tell you?

(ii) Apart from changing x to y and y to x, the answer is the same as for part (i).

Worksheet 15 (Solutions)

1. Predict, y, x.
2. Scatter, regression, straight line.
3. (c)
4. (b)
5. (b)
6. Did you beat 4000?
7. b = 11.92, a = 97.0. Predict (a) 335, (b) 395, (c) 455. The last of these is the least reliable, being furthest from the mean value of x, which is 22.9. Also 30 is just outside the range of temperature for these data.
8. b = 0.9766, a = 0.0836. For a Sonar reading of zero, predicted depth is 0.08. Using Minitab, and particularly the example in Section 15.7, we find that s_r = 0.1055 (see this result just above the ANOVA table in the Minitab output). A 95% confidence interval for the true depth when the Sonar reading of zero is −0.13 and +0.30, but since negative depths are impossible, quote 0 to 0.30.
9. b = 0.316, a = −2.60; (a) The regression equation cannot be used for a relative humidity of 0%, because it is well below the minimum value in the data; (b) When r.h. is 50%, predict moisture of 13.2; (c) When r.h. of 100%, again we cannot use the equation for a similar reason to part(a).
   Minitab gives s_r = 1.283 and *Calc F* = 42.24 and a p value of 0.001. Since *Tab F* = 6.61 and so *Calc F* > *Tab F*, reject the null hypothesis that the slope of the population regression line is zero. We could have reached the same conclusion by comparing the p value of 0.001 with 0.05.
10. Be careful which variable you choose to be the x and which the y variable; b = 0.339, a = 3.55; (a) 23.6 to 27.6 cals, (b) 29.5 to 31.8 cals, (c) 33.8 to 37.7.
11. b = −5.03, a = 101.7; s_r = 6.148 and the ANOVA table gives *Calc F* = 55.2, while *Tab F* = 5.32. So the slope is significantly different from zero. A 95% C.I. for β is −6.59 to −3.47.
12. b = 0.518, a = 54.4; (a) 67.3, 95% C.I. is 66.8 to 67.9, (b) 82.9, 95% C.I. is 82.2 to 83.5, (c) 98.4, 95% C.I. is 97.2 to 99.6, but part (c) is extrapolation since 85 is way beyond the maximum value in the data.
13. Scatter diagram shows no clear pattern, whereas we might have expected a negative correlation (the higher the IQ the shorter the time to complete). However, r = −0.066, which is negative but very small. The only reasonable conclusion from these data is that there may be several variables which affect the time to complete a crossword, but IQ doesn't appear to be one of them.

In this question, the correlation coefficient is very small and it is not at all obvious whether it is slightly positive or slightly negative. However, in Question 12, the correlation is very strong so the line 'drew itself'.

Worksheet 16 (Solutions)

1. *Calc* χ^2 = 0.47 *Tab* χ^2 = 7.82. Data consistent with theory.
2. (a) *Calc* χ^2 = 12.7 *Tab* χ^2 = 11.1. Reject uniform distribution.
   (b) *Calc* χ^2 = 60.4 *Tab* χ^2 = 11.1. Reject 2:1:5:4:5:3 distribution. Allowing for volume of traffic, significantly more accidents occur during the hours of darkness than expected.
3. *Calc* χ^2 = 13.7 *Tab* χ^2 = 5.99. Reject independence; more farms on flat land and fewer on mountainous land than expected.
4. *Calc* χ^2 = 5.5 *Tab* χ^2 = 5.99. Data consistent with 1:2:1 hypothesis.
5. (a) *Calc* χ^2 = 9.5 *Tab* χ^2 = 7.82. Data are not consistent with B (3, 0.5) distribution.
   (b) *Calc* χ^2 = 3.8 *Tab* χ^2 = 5.99. Data consistent with B(3, 0.4633) distribution, indicating significantly fewer boys than girls.
6. See Section 8.3 for grid method of selecting random points at which to place quadrats; count the number of pebbles within each quadrat, and choose ten using random number tables.
   Calc χ^2 = 1.9 *Tab* χ^2 = 12.6. Data consistent with B(10, 4) distribution.
7. *Calc* χ^2 = 5.5 *Tab* χ^2 = 11.1. Data consistent with B(20, 0.1) distribution.
8. See Section 8.3 for grid method for selecting 80 random points. The estimated value of m is 1.8. *Calc* χ^2 = 16.6 *Tab* χ^2 = 7.82. Data not consistent with random distribution.
9. *Calc* χ^2 = 3.9 *Tab* χ^2 = 9.49. Data consistent with a random distribution (m = 2).
10. (a) $\bar{x}$ = 2 s^2 = 4.13. Clearly the sample mean and the sample variance are not approximately equal, so it is not reasonable to assume that the Poisson is a good fit (model) here.
    (b) *Calc* χ^2 = 95.3 *Tab* χ^2 = 7.82. Data not consistent with a random distribution. There are many more cars than expected with either no defects or at least four defects.
11. (a) b = 25.11, *Calc W* = 0.973, *Tab W* = 0.842, data support normality.
    (b) b = 18.41, *Calc W* = 0.840, *Tab W* = 0.829, data support normality.

Appendix C

Statistical Tables

Table C.1 Cumulative Binomial Probabilities. The Table Gives the Probability of Obtaining r or Fewer Successes in n Independent Trials, Where p = Probability of Success in a Single Trial

n	r	$p=$ 0.01	0.02	0.03	0.04	0.05	0.06	0.07	0.08	0.09
n = 2	r = 0	.9801	.9604	.9409	.9216	.9025	.8836	.8649	.8464	.8281
	1	.9999	.9996	.9991	.9984	.9975	.9964	.9951	.9936	.9919
	2	1.0000	1.0000	1.0000	1.0000	1.0000	1.0000	1.0000	1.0000	1.0000
n = 5	r = 0	.9510	.9039	.8587	.8154	.7738	.7339	.6957	.6591	.6240
	1	.9990	.9962	.9915	.9852	.9774	.9681	.9575	.9456	.9326
	2	1.0000	.9999	.9997	.9994	.9988	.9980	.9969	.9955	.9937
	3		1.0000	1.0000	1.0000	1.0000	.9999	.9999	.9998	.9997
	4						1.0000	1.0000	1.0000	1.0000
n = 10	r = 0	.9044	.8171	.7374	.6648	.5987	.5386	.4840	.4344	.3894
	1	.9957	.9838	.9655	.9418	.9139	.8824	.8483	.8121	.7746
	2	.9999	.9991	.9972	.9938	.9885	.9812	.9717	.9599	.9460
	3	1.0000	1.0000	.9999	.9996	.9990	.9980	.9964	.9942	.9912
	4			1.0000	1.0000	.9999	.9999	.9997	.9994	.9990
	5					1.0000	1.0000	1.0000	1.0000	.9999
	6									1.0000
n = 20	r = 0	.8179	.6676	.5438	.4420	.3585	.2901	.2342	.1887	.1516
	1	.9831	.9401	.8802	.8103	.7358	.6605	.5869	.5169	.4516
	2	.9990	.9929	.9790	.9561	.9245	.8850	.8390	.7879	.7334
	3	1.0000	.9994	.9973	.9926	.9841	.9710	.9529	.9294	.9007
	4		1.0000	.9997	.9990	.9974	.9944	.9893	.9817	.9710

		0.10	0.15	0.20	0.25	0.30	0.35	0.40	0.45	0.50
	5			1.0000	.9999	.9997	.9991	.9981	.9962	.9932
	6				1.0000	1.0000	.9999	.9997	.9994	.9987
	7						1.0000	1.0000	.9999	.9998
	8								1.0000	1.0000
$n = 50$	$r = 0$	.6050	.3642	.2181	.1299	.0769	.0453	.0266	.0155	.0090
	1	.9106	.7358	.5553	.4005	.2794	.1900	.1265	.0827	.0532
	2	.9862	.9216	.8108	.6767	.5405	.4162	.3108	.2260	.1605
	3	.9984	.9822	.9372	.8609	.7604	.6473	.5327	.4253	.3303
	4	.9999	.9968	.9832	.9510	.8964	.8206	.7290	.6290	.5277
	5	1.0000	.9995	.9963	.9856	.9622	.9224	.8650	.7919	.7072
	6		.9999	.9993	.9964	.9882	.9711	.9417	.8981	.8404
	7		1.0000	.9999	.9992	.9968	.9906	.9780	.9562	.9232
	8			1.0000	.9999	.9992	.9973	.9927	.9833	.9672
	9				1.0000	.9998	.9993	.9978	.9944	.9875
	10					1.0000	.9998	.9994	.9983	.9957
	11						1.0000	.9999	.9995	.9987
	12							1.0000	.9999	.9996
	13								1.0000	.9999
	14									1.0000
$p =$		0.10	0.15	0.20	0.25	0.30	0.35	0.40	0.45	0.50
$n = 2$	$r = 0$	.8100	.7225	.6400	.5625	.4900	.4225	.3600	.3025	.2500
	1	.9900	.9775	.9600	.9375	.9100	.8775	.8400	.7975	.7500

(continued)

Table C.1 Cumulative Binomial Probabilities. The Table Gives the Probability of Obtaining r or Fewer Successes in n Independent Trials, Where p = Probability of Success in a Single Trial (Continued)

n	r	.10	.15	.20	.25	.30	.35	.40	.45	.50
	2	1.0000	1.0000	1.0000	1.0000	1.0000	1.0000	1.0000	1.0000	1.0000
5	0	.5905	.4437	.3277	.2373	.1681	.1160	.0778	.0503	.0313
	1	.9185	.8352	.7373	.6328	.5282	.4284	.3370	.2562	.1875
	2	.9914	.9734	.9421	.8965	.8369	.7648	.6826	.5831	.5000
	3	.9995	.9978	.9933	.9844	.9692	.9460	.9130	.8688	.8125
	4	1.0000	.9999	.9997	.9990	.9976	.9947	.9898	.9815	.9688
	5		1.0000	1.0000	1.0000	1.0000	1.0000	1.0000	1.0000	1.0000
10	0	.3487	.1969	.1074	.0563	.0282	.0135	.0060	.0025	.0010
	1	.7361	.5443	.3758	.2440	.1493	.0860	.0464	.0233	.0107
	2	.9298	.8202	.6778	.5256	.3828	.2616	.1673	.0996	.0547
	3	.9872	.9500	.8791	.7759	.6496	.5138	.3823	.2660	.1719
	4	.9984	.9901	.9672	.9219	.8497	.7515	.6331	.5044	.3770
	5	.9999	.9986	.9936	.9803	.9527	.9051	.8338	.7384	.6230
	6	1.0000	.9999	.9991	.9965	.9894	.9740	.9452	.8980	.8281
	7		1.0000	.9999	.9996	.9984	.9952	.9877	.9726	.9453
	8			1.0000	1.0000	.9999	.9995	.9983	.9955	.9893
	9					1.0000	1.0000	.9999	.9997	.9990
	10							1.0000	1.0000	1.0000
20	0	.1216	.0388	.0115	.0032	.0008	.0002			
	1	.3917	.1756	.0692	.0243	.0076	.0021	.0005	.0001	
	2	.6769	.4049	.2061	.0913	.0355	.0121	.0036	.0009	.0002

3	.8670	.6477	.4114	.2252	.1071	.0444	.0160	.0049	.0013
4	.9568	.8298	.6296	.4148	.2375	.1182	.0510	.0189	.0059
5	.9887	.9327	.8042	.6172	.4164	.2454	.1256	.0553	.0207
6	.9976	.9781	.9133	.7858	.6080	.4166	.2500	.1299	.0577
7	.9996	.9941	.9679	.8982	.7723	.6010	.4159	.2520	.1316
8	.9999	.9987	.9900	.9591	.8867	.7624	.5956	.4143	.2517
9	1.0000	.9998	.9974	.9861	.9520	.8782	.7553	.5914	.4119
10		1.0000	.9994	.9961	.9829	.9468	.8725	.7507	.5881
11			.9999	.9991	.9949	.9804	.9435	.8692	.7483
12			1.0000	.9998	.9987	.9940	.9790	.9420	.8684
13				1.0000	.9997	.9985	.9935	.9786	.9423
14					1.0000	.9997	.9984	.9936	.9793
15						1.0000	.9997	.9985	.9941
16							1.0000	.9997	.9987
17								1.0000	.9998
18									1.0000
$n = 50$ $r = 0$	.0052	.0003							
1	.0338	.0029	.0002						
2	.1117	.0142	.0013	.0001					
3	.2503	.0460	.0057	.0005					
4	.4312	.1121	.0185	.0021	.0002				

(continued)

Table C.1 Cumulative Binomial Probabilities. The Table Gives the Probability of Obtaining *r* or Fewer Successes in *n* Independent Trials, Where *p* = Probability of Success in a Single Trial (Continued)

r									
5	.6161	.2194	.0480	.0070	.0007	.0001			
6	.7702	.3613	.1034	.0194	.0025	.0002			
7	.8779	.5188	.1904	.0453	.0073	.0008	.0001		
8	.9421	.6681	.3073	.0916	.0183	.0025	.0002		
9	.9755	.7911	.4437	.1637	.0402	.0067	.0008	.0001	
10	.9906	.8801	.5836	.2622	.0789	.0160	.0022	.0002	
11	.9968	.9372	.7107	.3816	.1390	.0342	.0057	.0006	
12	.9990	.9699	.8139	.5110	.2229	.0661	.0133	.0018	.0002
13	.9997	.9868	.8894	.6370	.3279	.1163	.0280	.0045	.0005
14	.9999	.9947	.9393	.7481	.4468	.1878	.0540	.0104	.0013
15	1.0000	.9981	.9692	.8369	.5692	.2801	.0955	.0220	.0033
16		.9993	.9856	.9017	.6839	.3889	.1561	.0427	.0077
17		.9998	.9937	.9449	.7822	.5060	.2369	.0765	.0164
18		.9999	.9975	.9713	.8594	.6216	.3356	.1273	.0325
19		1.0000	.9991	.9861	.9152	.7264	.4465	.1974	.0595
20			.9997	.9937	.9522	.8139	.5610	.2862	.1013
21			.9999	.9974	.9749	.8813	.6701	.3900	.1611
22			1.0000	.9990	.9877	.9290	.7660	.5019	.2399
23				.9996	.9944	.9604	.8438	.6134	.3359
24				.9999	.9976	.9793	.9022	.7160	.4439

25	.5561	.8034	.9427	.9900	.9991	1.0000
26	.6641	.8721	.9686	.9955	.9997	
27	.7601	.9220	.9840	.9981	.9999	
28	.8389	.9556	.9924	.9993	1.0000	
29	.8987	.9765	.9966	.9997		
30	.9405	.9884	.9986	.9999		
31	.9675	.9947	.9995	1.0000		
32	.9836	.9978	.9998			
33	.9923	.9991	.9999			
34	.9967	.9997	1.0000			
35	.9987	.9999				
36	.9995	1.0000				
37	.9998					
38	1.0000					

Table C.2 Cumulative Poisson Probabilities. The Table Gives the Probability of r or Fewer Random Events Per Unit Time or Space, When the Average Number of Such Events is m

m =	0.1	0.2	0.3	0.4	0.5	0.6	0.7	0.8	0.9	1.0
r = 0	.9048	.8187	.7408	.6703	.6065	.5488	.4966	.4493	.4066	.3679
1	.9953	.9825	.9631	.9384	.9098	.8781	.8442	.8088	.7725	.7358
2	.9998	.9989	.9964	.9921	.9856	.9769	.9659	.9526	.9371	.9197
3	1.0000	.9999	.9997	.9992	.9982	.9966	.9942	.9909	.9865	.9810
4		1.0000	1.0000	.9999	.9998	.9996	.9992	.9986	.9977	.9963
5				1.0000	1.0000	1.0000	.9999	.9998	.9997	.9994
6							1.0000	1.0000	1.0000	.9999
7										1.0000

m =	1.1	1.2	1.3	1.4	1.5	1.6	1.7	1.8	1.9	2.0
r = 0	.3329	.3012	.2725	.2466	.2231	.2019	.1827	.1653	.1496	.1353
1	.6990	.6626	.6268	.5918	.5578	.5249	.4932	.4628	.4337	.4060
2	.9004	.8795	.8571	.8335	.8088	.7834	.7572	.7306	.7037	.6767
3	.9743	.9662	.9569	.9463	.9344	.9212	.9068	.8913	.8747	.8571
4	.9946	.9923	.9893	.9857	.9814	.9763	.9704	.9636	.9559	.9473
5	.9990	.9985	.9978	.9968	.9955	.9940	.9920	.9896	.9868	.9834
6	.9999	.9997	.9996	.9994	.9991	.9987	.9981	.9974	.9966	.9955
7	1.0000	1.0000	.9999	.9999	.9998	.9997	.9996	.9994	.9992	.9989
8			1.0000	1.0000	1.0000	1.0000	.9999	.9999	.9998	.9998
9							1.0000	1.0000	1.0000	1.0000

$m =$	2.1	2.2	2.3	2.4	2.5	2.6	2.7	2.8	2.9	3.0
$r = 0$	.1225	.1108	.1003	.0907	.0821	.0743	.0672	.0608	.0550	.0498
1	.3796	.3546	.3309	.3084	.2873	.2674	.2487	.2311	.2146	.1991
2	.6496	.6227	.5960	.5697	.5438	.5184	.4936	.4695	.4460	.4232
3	.8386	.8194	.7993	.7787	.7576	.7360	.7141	.6919	.6696	.6472
4	.9379	.9275	.9162	.9041	.8912	.8774	.8629	.8477	.8318	.8153
5	.9796	.9751	.9700	.9643	.9580	.9510	.9433	.9349	.9258	.9161
6	.9941	.9925	.9906	.9884	.9858	.9828	.9794	.9756	.9713	.9665
7	.9985	.9980	.9974	.9967	.9958	.9947	.9934	.9919	.9901	.9881
8	.9997	.9995	.9994	.9991	.9989	.9985	.9981	.9976	.9969	.9962
9	.9999	.9999	.9999	.9998	.9997	.9996	.9995	.9993	.9991	.9989
10	1.0000	1.0000	1.0000	1.0000	.9999	.9999	.9999	.9998	.9998	.9997
11					1.0000	1.0000	1.0000	1.0000	.9999	.9999
12									1.0000	1.0000

$m =$	3.1	3.2	3.3	3.4	3.5	3.6	3.7	3.8	3.9	4.0
$r = 0$	.0450	.0408	.0369	.0334	.0302	.0273	.0247	.0224	.0202	.0183
1	.1847	.1712	.1586	.1468	.1359	.1257	.1162	.1074	.0992	.0916
2	.4012	.3799	.3594	.3397	.3208	.3027	.2854	.2689	.2531	.2381
3	.6248	.6025	.5803	.5584	.5366	.5152	.4942	.4735	.4532	.4335
4	.7982	.7806	.7626	.7442	.7254	.7064	.6872	.6678	.6484	.6288
5	.9057	.8946	.8829	.8705	.8576	.8441	.8301	.8156	.8006	.7851
6	.9612	.9554	.9490	.9421	.9347	.9267	.9182	.9091	.8995	.8893
7	.9858	.9832	.9802	.9769	.9733	.9692	.9648	.9599	.9546	.9489

(continued)

Table C.2 Cumulative Poisson Probabilities. The Table Gives the Probability of r or Fewer Random Events Per Unit Time or Space, When the Average Number of Such Events is m (Continued)

m =	3.1	3.2	3.3	3.4	3.5	3.6	3.7	3.8	3.9	4.0
8	.9953	.9943	.9931	.9917	.9901	.9883	.9863	.9840	.9815	.9786
9	.9986	.9982	.9978	.9973	.9967	.9960	.9952	.9942	.9931	.9919
10	.9996	.9995	.9994	.9992	.9990	.9987	.9984	.9981	.9977	.9972
11	.9999	.9999	.9998	.9998	.9997	.9996	.9995	.9994	.9993	.9991
12	1.0000	1.0000	1.0000	.9999	.9999	.9999	.9999	.9998	.9998	.9997
13				1.0000	1.0000	1.0000	1.0000	1.0000	.9999	.9999
14									1.0000	1.0000

m =	4.1	4.2	4.3	4.4	4.5	4.6	4.7	4.8	4.9	5.0
r = 0	.0166	.0150	.0136	.0123	.0111	.0101	.0091	.0082	.0074	.0067
1	.0845	.0780	.0719	.0663	.0611	.0563	.0518	.0477	.0439	.0404
2	.2238	.2102	.1974	.1851	.1736	.1626	.1523	.1425	.1333	.1247
3	.4142	.3954	.3772	.3594	.3423	.3257	.3097	.2942	.2793	.2650
4	.6093	.5898	.5704	.5512	.5321	.5132	.4946	.4763	.4582	.4405
5	.7693	.7531	.7367	.7199	.7029	.6858	.6684	.6510	.6335	.6160
6	.8786	.8675	.8558	.8436	.8311	.8180	.8046	.7908	.7767	.7622
7	.9427	.9361	.9290	.9214	.9134	.9049	.8960	.8867	.8769	.8666
8	.9755	.9721	.9683	.9642	.9597	.9549	.9497	.9442	.9382	.9319
9	.9905	.9889	.9871	.9851	.9829	.9805	.9778	.9749	.9717	.9682
10	.9966	.9959	.9952	.9943	.9933	.9922	.9910	.9896	.9880	.9863
11	.9989	.9986	.9983	.9980	.9976	.9971	.9966	.9960	.9953	.9945

r										
12	.9997	.9996	.9995	.9993	.9992	.9990	.9988	.9986	.9983	.9980
13	.9999	.9999	.9998	.9998	.9997	.9997	.9996	.9995	.9994	.9993
14	1.0000	1.0000	1.0000	.9999	.9999	.9999	.9999	.9999	.9998	.9998
15				1.0000	1.0000	1.0000	1.0000	1.0000	.9999	.9999
16									1.0000	1.0000

m =	5.2	5.4	5.6	5.8	6.0	6.2	6.4	6.6	6.8	7.0
r = 0	.0055	.0045	.0037	.0030	.0025	.0020	.0017	.0014	.0011	.0009
1	.0342	.0289	.0244	.0206	.0174	.0146	.0123	.0103	.0087	.0073
2	.1088	.0948	.0824	.0715	.0620	.0536	.0463	.0400	.0344	.0296
3	.2381	.2133	.1906	.1700	.1512	.1342	.1189	.1052	.0928	.0818
4	.4061	.3733	.3422	.3127	.2851	.2592	.2351	.2127	.1920	.1730
5	.5809	.5461	.5119	.4783	.4457	.4141	.3837	.3547	.3270	.3007
6	.7324	.7017	.6703	.6384	.6063	.5742	.5423	.5108	.4799	.4497
7	.8449	.8217	.7970	.7710	.7440	.7160	.6873	.6581	.6285	.5987
8	.9181	.9027	.8857	.8672	.8472	.8259	.8033	.7796	.7548	.7291
9	.9603	.9512	.9409	.9292	.9161	.9016	.8858	.8686	.8502	.8305
10	.9823	.9775	.9718	.9651	.9574	.9486	.9386	.9274	.9151	.9015
11	.9927	.9904	.9875	.9841	.9799	.9750	.9693	.9627	.9552	.9467
12	.9972	.9962	.9949	.9932	.9912	.9887	.9857	.9821	.9779	.9730
13	.9990	.9986	.9980	.9973	.9964	.9952	.9937	.9920	.9898	.9872
14	.9997	.9995	.9993	.9990	.9986	.9981	.9974	.9966	.9956	.9943
15	.9999	.9998	.9998	.9996	.9995	.9993	.9990	.9986	.9982	.9976
16	1.0000	.9999	.9999	.9999	.9998	.9997	.9996	.9995	.9993	.9990
17		1.0000	1.0000	1.0000	.9999	.9999	.9999	.9998	.9997	.9996

(continued)

Table C.2 Cumulative Poisson Probabilities. The Table Gives the Probability of r or Fewer Random Events Per Unit Time or Space, When the Average Number of Such Events is m (Continued)

m =	5.2	5.4	5.6	5.8	6.0	6.2	6.4	6.6	6.8	7.0
18					1.0000	1.0000	1.0000	.9999	.9999	.9999
19								1.0000	1.0000	1.0000

m =	7.2	7.4	7.6	7.8	8.0	8.2	8.4	8.6	8.8	9.0
r = 0	.0007	.0006	.0005	.0004	.0003	.0003	.0002	.0002	.0002	.0001
1	.0061	.0051	.0043	.0036	.0030	.0025	.0021	.0018	.0015	.0012
2	.0255	.0219	.0188	.0161	.0138	.0118	.0100	.0086	.0073	.0062
3	.0719	.0632	.0554	.0485	.0424	.0370	.0323	.0281	.0244	.0212
4	.1555	.1395	.1249	.1117	.0996	.0887	.0789	.0701	.0621	.0550
5	.2759	.2526	.2307	.2103	.1912	.1736	.1573	.1422	.1284	.1157
6	.4204	.3920	.3646	.3384	.3134	.2896	.2670	.2457	.2256	.2068
7	.5689	.5393	.5100	.4812	.4530	.4254	.3987	.3728	.3478	.3239
8	.7027	.6757	.6482	.6204	.5925	.5647	.5369	.5094	.4823	.4557
9	.8096	.7877	.7649	.7411	.7166	.6915	.6659	.6400	.6137	.5874
10	.8867	.8707	.8535	.8352	.8159	.7955	.7743	.7522	.7294	.7060
11	.9371	.9265	.9148	.9020	.8881	.8731	.8571	.8400	.8220	.8030
12	.9673	.9609	.9536	.9454	.9362	.9261	.9150	.9029	.8898	.8758
13	.9841	.9805	.9762	.9714	.9658	.9595	.9524	.9445	.9358	.9261
14	.9927	.9908	.9886	.9859	.9827	.9791	.9749	.9701	.9647	.9585
15	.9969	.9959	.9948	.9934	.9918	.9898	.9875	.9848	.9816	.9780
16	.9987	.9983	.9978	.9971	.9963	.9953	.9941	.9926	.9909	.9889
17	.9995	.9993	.9991	.9988	.9984	.9979	.9973	.9966	.9957	.9947

	9.2	9.4	9.6	9.8	10.0	11.0	12.0	13.0	14.0	15.0
18	.9998	.9997	.9996	.9995	.9993	.9991	.9989	.9985	.9981	.9976
19	.9999	.9999	.9999	.9998	.9997	.9997	.9995	.9994	.9992	.9989
20	1.0000	1.0000	1.0000	.9999	.9999	.9999	.9998	.9998	.9997	.9996
21				1.0000	1.0000	1.0000	.9999	.9999	.9999	.9998
22							1.0000	1.0000	1.0000	.9999
23										1.0000

$m =$	9.2	9.4	9.6	9.8	10.0	11.0	12.0	13.0	14.0	15.0
$r = 0$	.0001	.0001	.0001	.0001						
1	.0010	.0009	.0007	.0006	.0005	.0002	.0001			
2	.0053	.0045	.0038	.0033	.0028	.0012	.0005	.0002	.0001	
3	.0184	.0160	.0138	.0120	.0103	.0049	.0023	.0011	.0005	.0002
4	.0486	.0429	.0378	.0333	.0293	.0151	.0076	.0037	.0018	.0009
5	.1041	.0935	.0838	.0750	.0671	.0375	.0203	.0107	.0055	.0028
6	.1892	.1727	.1574	.1433	.1301	.0786	.0458	.0259	.0142	.0076
7	.3010	.2792	.2584	.2388	.2202	.1432	.0895	.0540	.0316	.0180
8	.4296	.4042	.3796	.3558	.3328	.2320	.1550	.0998	.0621	.0374
9	.5611	.5349	.5089	.4832	.4579	.3405	.2424	.1658	.1094	.0699
10	.6820	.6576	.6329	.6080	.5830	.4599	.3472	.2517	.1757	.1185
11	.7832	.7626	.7412	.7193	.6968	.5793	.4616	.3532	.2600	.1848
12	.8607	.8448	.8279	.8101	.7916	.6887	.5760	.4631	.3585	.2676
13	.9156	.9042	.8919	.8786	.8645	.7813	.6815	.5730	.4644	.3632
14	.9517	.9441	.9357	.9265	.9165	.8540	.7720	.6751	.5704	.4657
15	.9738	.9691	.9638	.9579	.9513	.9074	.8444	.7636	.6694	.5681

(continued)

Table C.2 Cumulative Poisson Probabilities. The Table Gives the Probability of *r* or Fewer Random Events Per Unit Time or Space, When the Average Number of Such Events is *m* (Continued)

$m =$	9.2	9.4	9.6	9.8	10.0	11.0	12.0	13.0	14.0	15.0
16	.9865	.9838	.9806	.9770	.9730	.9441	.8987	.8355	.7559	.6641
17	.9934	.9919	.9902	.9881	.9857	.9678	.9370	.8905	.8272	.7489
18	.9969	.9962	.9952	.9941	.9928	.9823	.9626	.9302	.8826	.8195
19	.9986	.9983	.9978	.9972	.9965	.9907	.9787	.9573	.9235	.8752
20	.9994	.9992	.9990	.9987	.9984	.9953	.9884	.9750	.9521	.9170
21	.9998	.9997	.9996	.9995	.9993	.9977	.9939	.9859	.9712	.9469
22	.9999	.9999	.9998	.9998	.9997	.9990	.9970	.9924	.9833	.9673
23	1.0000	.9999	.9999	.9999	.9999	.9995	.9985	.9960	.9907	.9805
24		1.0000	1.0000	1.0000	1.0000	.9998	.9993	.9980	.9950	.9888
25						.9999	.9997	.9990	.9974	.9938
26						1.0000	.9999	.9995	.9987	.9967
27							.9999	.9998	.9994	.9983
28							1.0000	.9999	.9997	.9991
29								1.0000	.9999	.9996
30									.9999	.9998
31									1.0000	.9999
32										1.0000

Table C.3(a) Normal Distribution Function for a Normal Distribution with a Mean, μ, and Standard Deviation, σ, and a Particular Value of x, Calculate $z = (x - \mu)/\sigma$. The Table Gives the Area to the Left of x, see Fig. C.1

$z = \dfrac{(x - \mu)}{\sigma}$	0.00	0.01	0.02	0.03	0.04	0.05	0.06	0.07	0.08	0.09
0.0	.5000	.5040	.5080	.5120	.5160	.5199	.5239	.5279	.5319	.5359
0.1	.5398	.5438	.5478	.5517	.5557	.5596	.5636	.5675	.5714	.5753
0.2	.5793	.5832	.5871	.5910	.5948	.5987	.6026	.6064	.6103	.6141
0.3	.6179	.6217	.6255	.6293	.6331	.6368	.6406	.6443	.6480	.6517
0.4	.6554	.6591	.6628	.6664	.6700	.6736	.6772	.6808	.6844	.6879
0.5	.6915	.6950	.6985	.7019	.7054	.7088	.7123	.7157	.7190	.7224
0.6	.7257	.7291	.7324	.7357	.7389	.7422	.7454	.7486	.7517	.7549
0.7	.7580	.7611	.7642	.7673	.7704	.7734	.7764	.7794	.7823	.7852
0.8	.7881	.7910	.7939	.7967	.7995	.8023	.8051	.8078	.8106	.8133
0.9	.8159	.8186	.8212	.8238	.8264	.8289	.8315	.8340	.8365	.8389
1.0	.8413	.8438	.8461	.8485	.8508	.8531	.8554	.8577	.8599	.8621
1.1	.8643	.8665	.8686	.8708	.8729	.8749	.8770	.8790	.8810	.8830
1.2	.8849	.8869	.8888	.8907	.8925	.8944	.8962	.8980	.8997	.9015
1.3	.9032	.9049	.9066	.9082	.9099	.9115	.9131	.9147	.9162	.9177
1.4	.9192	.9207	.9222	.9236	.9251	.9265	.9279	.9292	.9306	.9319
1.5	.9332	.9345	.9357	.9370	.9382	.9394	.9406	.9418	.9429	.9441
1.6	.9452	.9463	.9474	.9484	.9495	.9505	.9515	.9525	.9535	.9545
1.7	.9554	.9564	.9573	.9582	.9591	.9599	.9608	.9616	.9625	.9633
1.8	.9641	.9649	.9656	.9664	.9671	.9678	.9686	.9693	.9699	.9706
1.9	.9713	.9719	.9726	.9732	.9738	.9744	.9750	.9756	.9761	.9767
2.0	.9772	.9778	.9783	.9788	.9793	.9798	.9803	.9808	.9812	.9817

(continued)

Table C.3(a) Normal Distribution Function for a Normal Distribution with a Mean, μ, and Standard Deviation, σ, and a Particular Value of x, Calculate $z = (x - \mu)/\sigma$. The Table Gives the Area to the Left of x, see Fig. C.1 (Continued)

z										
2.1	.9821	.9826	.9830	.9834	.9838	.9842	.9846	.9850	.9854	.9857
2.2	.9861	.9864	.9868	.9871	.9875	.9878	.9881	.9884	.9887	.9890
2.3	.9893	.9896	.9898	.9901	.9904	.9906	.9909	.9911	.9913	.9916
2.4	.9918	.9920	.9922	.9925	.9927	.9929	.9931	.9932	.9934	.9936
2.5	.9938	.9940	.9941	.9943	.9945	.9946	.9948	.9949	.9951	.9952
2.6	.9953	.9955	.9956	.9957	.9959	.9960	.9961	.9962	.9963	.9964
2.7	.9965	.9966	.9967	.9968	.9969	.9970	.9971	.9972	.9973	.9974
2.8	.9974	.9975	.9976	.9977	.9977	.9978	.9979	.9979	.9980	.9981
2.9	.9981	.9982	.9982	.9983	.9984	.9984	.9985	.9985	.9986	.9986
3.0	.9987	.9987	.9987	.9988	.9988	.9989	.9989	.9989	.9990	.9990
3.1	.9990	.9991	.9991	.9991	.9992	.9992	.9992	.9992	.9993	.9993
3.2	.9993	.9993	.9994	.9994	.9994	.9994	.9994	.9995	.9995	.9995
3.3	.9995	.9995	.9995	.9996	.9996	.9996	.9996	.9996	.9996	.9997
3.4	.9997	.9997	.9997	.9997	.9997	.9997	.9997	.9997	.9997	.9998

Table C.3(b) Upper Percentage Points for the Normal Distribution. The Table Gives the Values of z for Various Right-Hand Tail Areas, α, See Fig. C.2.

α	0.05	0.025	0.01	0.005	0.001	0.0005
z	1.645	1.96	2.33	2.58	3.09	3.29

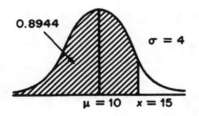

Figure C.1 $\quad z = \dfrac{x - \mu}{\sigma} = \dfrac{15 - 10}{4} = 1.25$

Figure C.2 Right-Hand Tail Area of a Normal Distribution

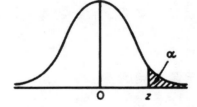

Table C.4 Random Numbers

30 89 34 43 98	38 51 15 30 26	02 57 93 32 67	19 91 72 23 06	59 24 11 06 50
79 50 49 98 07	05 88 29 05 29	73 15 65 17 92	26 05 21 60 73	55 48 97 54 50
53 64 54 20 36	05 26 90 12 98	73 98 56 47 60	44 54 45 97 21	25 70 96 58 72
87 23 75 21 50	54 47 46 35 72	11 66 30 44 63	69 50 82 74 58	98 25 68 47 79
91 54 58 41 48	70 11 94 79 12	36 63 12 52 72	43 41 11 52 98	91 77 91 85 00
92 41 24 08 42	64 96 82 07 01	40 00 95 09 30	23 40 08 19 78	55 50 92 84 96
65 63 25 34 62	93 01 96 23 23	81 31 94 09 02	75 98 27 85 59	53 09 94 37 37
93 64 13 39 70	98 38 71 77 89	47 98 47 22 09	98 85 91 86 42	30 60 34 07 23
92 44 97 54 10	53 06 50 66 76	13 89 09 41 28	93 04 75 68 09	78 22 82 88 10
69 37 57 14 85	43 72 12 89 80	07 01 17 91 30	17 00 49 53 99	46 51 26 74 28
88 13 45 79 30	32 44 38 84 94	26 65 83 04 43	88 70 99 09 89	31 59 08 29 11
30 86 16 00 13	89 22 16 01 29	98 65 92 13 36	26 88 58 18 89	67 19 71 92 28
19 39 94 95 22	70 99 77 50 29	30 16 69 87 18	48 56 34 92 85	42 54 25 72 84
04 01 90 59 21	33 16 80 53 51	90 02 92 76 72	03 82 77 75 72	33 44 87 58 29
17 45 23 69 94	53 68 59 13 13	68 39 80 62 31	70 44 32 01 47	54 43 70 97 08
13 35 10 58 52	66 73 38 05 80	45 71 76 21 80	10 58 72 17 06	50 72 97 41 48
07 48 12 02 82	51 55 21 61 13	44 27 63 97 04	56 13 88 48 02	34 15 84 30 87
08 16 12 72 05	72 10 63 76 44	92 84 98 81 43	71 66 24 27 16	06 32 39 21 89
51 94 42 32 70	21 82 38 94 46	59 34 75 61 97	72 76 50 50 30	70 27 08 16 72
06 78 72 46 93	36 77 57 19 49	99 18 26 11 63	74 29 96 14 57	76 72 92 86 28
39 14 12 52 96	24 33 70 06 77	56 59 42 11 80	33 05 63 40 14	22 70 62 17 05
71 31 34 02 97	98 57 79 44 68	06 62 74 23 69	77 41 05 17 26	41 68 37 19 53
57 64 15 72 66	13 41 98 06 19	64 53 36 19 16	19 90 71 70 74	04 03 30 05 34
64 26 20 69 40	12 85 65 75 73	92 57 43 97 70	71 28 02 89 91	86 98 64 56 73
91 38 37 54 09	99 35 01 78 03	09 53 57 79 53	50 23 00 90 49	45 28 45 00 94

95	79	70	14	29	39	53	53	55	85	70	66	96	59	96	90	41	25	77	28
47	19	09	18	04	31	00	86	43	46	42	72	09	30	59	81	42	42	06	48
02	26	75	93	05	57	25	89	89	57	44	53	20	78	54	69	38	59	95	34
84	70	64	03	54	03	63	24	95	64	13	13	42	32	71	79	59	94	57	98
20	17	53	34	88	12	35	26	51	62	81	80	49	72	56	06	13	46	40	19
48	17	89	62	19	07	05	12	66	26	48	74	79	14	92	21	57	66	07	71
94	23	72	93	07	89	69	72	79	75	69	42	70	34	96	73	62	82	34	26
08	61	14	56	84	60	45	24	33	18	41	19	73	50	30	63	97	49	89	46
18	23	06	03	14	97	81	14	62	93	67	47	49	05	08	74	79	58	27	62
32	02	28	65	24	36	19	98	51	50	03	30	92	78	14	73	55	51	65	48
36	65	73	14	96	25	53	18	20	60	56	18	52	61	23	09	89	85	84	93
19	58	22	34	11	88	61	48	62	93	42	77	06	39	51	78	09	80	87	79
41	27	50	79	04	21	04	25	37	03	41	84	75	14	26	14	04	98	32	99
30	95	39	37	46	43	03	21	23	31	58	11	44	07	84	26	14	81	12	82
07	72	47	50	83	95	91	92	41	20	63	66	18	33	51	95	20	29	01	46
64	38	61	50	68	39	32	26	32	18	19	58	58	15	40	63	01	35	47	30
32	10	37	22	37	89	21	07	30	31	89	47	10	99	57	41	77	21	51	85
50	91	61	77	73	53	38	73	17	22	65	26	88	51	00	71	60	58	15	61
17	93	14	22	52	65	30	19	39	82	34	85	26	68	47	60	42	37	01	10
22	43	05	35	93	52	18	52	16	31	85	01	04	89	51	43	77	77	72	13
07	27	51	65	14	91	31	84	54	72	30	28	35	37	50	23	21	73	09	30
54	44	21	09	86	18	27	03	61	95	48	91	30	81	78	53	10	82	79	98
45	03	30	43	76	08	53	75	94	27	50	85	87	45	69	26	07	13	16	93
29	31	45	61	75	91	61	29	17	51	12	25	09	08	74	08	60	07	04	44
89	81	05	03	82	62	99	44	51	87	58	78	97	69	82	71	17	90	14	42

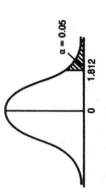

Figure C.3 *t* Distribution With ν = 10 d.f.

Table C.5 Percentage Points of the *t* Distribution. For a *t* Distribution With ν Degrees of Freedom, the Table Gives the Values of *t* which are exceeded with Probability α. Figure C.3 Shows a *t* Distribution with ν = 10 d.f.

α =	0.10	0.05	0.025	0.01	0.005	0.001	0.0005
ν = 1	3.076	6.314	12.706	31.821	63.657	318.31	636.62
2	1.886	2.920	4.303	6.965	9.925	22.326	31.598
3	1.638	2.353	3.182	4.541	5.841	10.213	12.924
4	1.533	2.132	2.776	3.747	4.604	7.173	8.610
5	1.476	2.015	2.571	3.365	4.032	5.893	6.869
6	1.440	1.943	2.447	3.143	3.707	5.208	5.959
7	1.415	1.895	2.365	2.998	3.499	4.785	5.408
8	1.397	1.860	2.306	2.896	3.355	4.501	5.041
9	1.383	1.833	2.262	2.821	3.250	4.297	4.781
10	1.372	1.812	2.228	2.764	3.169	4.144	4.587
11	1.363	1.796	2.201	2.718	3.106	4.025	4.437
12	1.356	1.782	2.179	2.681	3.055	3.930	4.318
13	1.350	1.771	2.160	2.650	3.012	3.852	4.221
14	1.345	1.761	2.145	2.624	2.977	3.787	4.140

15	1.341	1.753	2.131	2.602	2.947	3.733	4.073
16	1.337	1.746	2.120	2.583	2.921	3.686	4.015
17	1.333	1.740	2.110	2.567	2.898	3.646	3.965
18	1.330	1.734	2.101	2.552	2.878	3.610	3.922
19	1.328	1.729	2.093	2.539	2.861	3.579	3.883
20	1.325	1.725	2.086	2.528	2.845	3.552	3.850
21	1.323	1.721	2.080	2.518	2.831	3.527	3.819
22	1.321	1.717	2.074	2.508	2.819	3.505	3.792
23	1.319	1.714	2.069	2.500	2.807	3.485	3.767
24	1.318	1.711	2.064	2.492	2.797	3.467	3.745
25	1.316	1.708	2.060	2.485	2.787	3.450	3.725
26	1.315	1.706	2.056	2.479	2.779	3.435	3.707
27	1.314	1.703	2.052	2.473	2.771	3.421	3.690
28	1.313	1.701	2.048	2.467	2.763	3.408	3.674
29	1.311	1.699	2.045	2.462	2.756	3.396	3.659
30	1.310	1.697	2.042	2.457	2.750	3.385	3.646
40	1.303	1.684	2.021	2.423	2.704	3.307	3.551
60	1.296	1.671	2.000	2.390	2.660	3.232	3.460
120	1.289	1.658	1.980	2.358	2.617	3.160	3.373
∞	1.282	1.645	1.960	2.326	2.576	3.090	3.291

Table C.6 5% Points of the F Distribution. The Tabulated Value is $F_{0.05,\ \nu_1,\ \nu_2}$, Where $P(X > F_{0.05,\ \nu_1,\ \nu_2}) = 0.05$ When X has the F-distribution With ν_1, ν_2 Degrees of Freedom. The 95% Point may be Obtained Using $F'_{0.95,\ \nu_1,\ \nu_2} = \dfrac{1}{F_{0.05,\ \nu_1,\ \nu_2}}$

e.g. $F_{0.95,\ 12,\ 8} = \dfrac{1}{F_{0.05,\ 8,\ 12}} = \dfrac{1}{2.85} = 0.351$

$\nu_1 =$	1	2	3	4	5	6	7	8	10	12	24	∞
$\nu_2 = 1$	161.4	199.5	215.7	224.6	230.2	234.0	236.8	238.9	241.9	243.9	249.1	254.3
2	18.5	19.0	19.2	19.2	19.3	19.3	19.4	19.4	19.4	19.4	19.5	19.5
3	10.1	9.55	9.28	9.12	9.01	8.94	8.89	8.85	8.79	8.74	8.64	8.53
4	7.71	6.94	6.59	6.39	6.26	6.16	6.09	6.04	5.96	5.91	5.77	5.63
5	6.61	5.79	5.41	5.19	5.05	4.95	4.88	4.82	4.74	4.68	4.53	4.36
6	5.99	5.14	4.76	4.53	4.39	4.28	4.21	4.15	4.06	4.00	3.84	3.67
7	5.59	4.74	4.35	4.12	3.97	3.87	3.79	3.73	3.64	3.57	3.41	3.23
8	5.32	4.46	4.07	3.84	3.69	3.58	3.50	3.44	3.35	3.28	3.12	2.93
9	5.12	4.26	3.86	3.63	3.48	3.37	3.29	3.23	3.14	3.07	2.90	2.71
10	4.96	4.10	3.71	3.48	3.33	3.22	3.14	3.07	2.98	2.91	2.74	2.54
12	4.75	3.89	3.49	3.26	3.11	3.00	2.91	2.85	2.75	2.69	2.51	2.30
15	4.54	3.68	3.29	3.06	2.90	2.79	2.71	2.64	2.54	2.48	2.29	2.07
20	4.35	3.49	3.10	2.87	2.71	2.60	2.51	2.45	2.35	2.28	2.08	1.84
24	4.26	3.40	3.01	2.78	2.62	2.51	2.42	2.36	2.25	2.18	1.98	1.73
30	4.17	3.32	2.92	2.69	2.53	2.42	2.33	2.27	2.16	2.09	1.89	1.62
40	4.08	3.23	2.84	2.61	2.45	2.34	2.25	2.18	2.08	2.00	1.79	1.51
60	4.00	3.15	2.76	2.53	2.37	2.25	2.17	2.10	1.99	1.92	1.70	1.39
∞	3.84	3.00	2.60	2.37	2.21	2.10	2.01	1.94	1.83	1.75	1.52	1.00

Figure C.4 *F* Distribution With 8, 12 d.f.

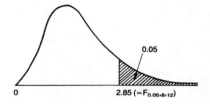

Table C.7   Values of *T* for the Wilcoxon Signed Rank Test

	Level of Significance for One-Sided H_1			
	0.05	0.25	0.01	0.005
	Level of Significance for Two-Sided H_1			
n	0.10	0.05	0.02	0.01
5	0	—	—	—
6	2	0	—	—
7	3	2	0	—
8	5	3	1	0
9	8	5	3	1
10	10	8	5	3
11	13	10	7	5
12	17	13	9	7
13	21	17	12	9
14	25	21	15	12
15	30	25	19	15
16	35	29	23	19
17	41	34	27	23
18	47	40	32	27
19	53	46	37	32
20	60	52	43	37
21	67	58	49	42
22	75	65	55	48
23	83	73	62	54
24	91	81	69	61
25	100	89	76	68

Table C.8 Values of U for the Mann-Whitney U Test Critical Values of U for the Mann-Whitney Test for 0.05 (First Value) and 0.01 (Second Value) Significance Levels for Two-Sided H_1, and for 0.025 and 0.005 Levels for One-Sided H_1.

Each cell shows two values: the first (upper) value is for the 0.05 / 0.025 level, the second (lower) value is for the 0.01 / 0.005 level (shown here as *first / second*). "—" indicates no critical value.

$n_1 \backslash n_2$	1	2	3	4	5	6	7	8	9	10	11	12	13	14	15	16	17	18	19	20
1	—	—	—	—	—	—	—	—	—	—	—	—	—	—	—	—	—	—	—	—
2	—	—	—	—	—	—	—	0/—	0/—	0/—	0/—	1/—	1/—	1/—	1/—	1/—	2/—	2/—	2/0	2/0
3	—	—	—	—	0/—	1/—	1/—	2/—	2/0	3/0	3/0	4/1	4/1	5/1	5/2	6/2	6/2	7/2	7/3	8/3
4	—	—	—	0/—	1/—	2/0	3/0	4/1	4/1	5/2	6/2	7/3	8/3	9/4	10/5	11/5	11/6	12/6	13/7	14/8
5	—	—	0/—	1/—	2/0	3/1	5/1	6/2	7/3	8/4	9/5	11/6	12/7	13/7	14/8	15/9	17/10	18/11	19/12	20/13
6	—	—	1/—	2/0	3/1	5/2	6/3	8/4	10/5	11/6	13/7	14/9	16/10	17/11	19/12	21/13	22/15	24/16	25/17	27/18
7	—	—	1/—	3/0	5/1	6/3	8/4	10/6	12/7	14/9	16/10	18/12	20/13	22/15	24/16	26/18	28/19	30/21	32/22	34/24
8	—	0/—	2/—	4/1	6/2	8/4	10/6	13/7	15/9	17/11	19/13	22/15	24/17	26/18	29/20	31/22	34/24	36/26	38/28	41/30
9	—	0/—	2/0	4/1	7/3	10/5	12/7	15/9	17/11	20/13	23/16	26/18	28/20	31/22	34/24	37/27	39/29	42/31	45/33	48/36
10	—	0/—	3/0	5/2	8/4	11/6	14/9	17/11	20/13	23/16	26/18	29/21	33/24	36/26	39/29	42/31	45/34	48/37	52/39	55/42
11	—	0/—	3/0	6/2	9/5	13/7	16/10	19/13	23/16	26/18	30/21	33/24	37/27	40/30	44/33	47/36	51/39	55/42	58/45	62/48

12	—	1	4	7	11	14	18	22	26	29	33	37	41	45	49	53	57	61	65	69
	—	—	1	3	6	9	12	15	18	21	24	27	31	34	37	41	44	47	51	54
13	—	1	4	8	12	16	20	24	28	33	37	41	45	50	54	59	63	67	72	76
	—	—	1	3	7	10	13	17	20	24	27	31	34	38	42	45	49	53	57	60
14	—	1	5	9	13	17	22	26	31	36	40	45	50	55	59	64	69	74	78	83
	—	—	1	4	7	11	15	18	22	26	30	34	38	42	46	50	54	58	63	67
15	—	1	5	10	14	19	24	29	34	39	44	49	54	59	64	70	75	80	85	90
	—	—	2	5	8	12	16	20	24	29	33	37	42	46	51	55	60	64	69	73
16	—	1	6	11	15	21	26	31	37	42	47	53	59	64	70	75	81	86	92	98
	—	—	2	5	9	13	18	22	27	31	36	41	45	50	55	60	65	70	74	79
17	—	2	6	11	17	22	28	34	39	45	51	57	63	69	75	81	87	93	99	105
	—	—	2	6	10	15	19	24	29	34	39	44	49	54	60	65	70	75	81	86
18	—	2	7	12	18	24	30	36	42	48	55	61	67	74	80	86	93	99	106	112
	—	—	2	6	11	16	21	26	31	37	42	47	53	58	64	70	75	81	87	92
19	—	2	7	13	19	25	32	38	45	52	58	65	72	78	85	92	99	106	113	119
	—	0	3	7	12	17	22	28	33	39	45	51	57	63	69	74	81	87	93	99
20	—	2	8	14	20	27	34	41	48	55	62	69	76	83	90	98	105	112	119	127
	—	0	3	8	13	18	24	30	36	42	48	54	60	67	73	79	86	92	99	105

Table C.9 Critical Values of the Studentised Range Statistic, q

Residual d.f.	5% Level of Significance			
	$p = 2$	$p = 3$	$p = 4$	$p = 5$
1	17.97	26.98	32.82	37.08
2	6.085	8.331	9.798	10.88
3	4.501	5.910	6.825	7.502
4	3.927	5.040	5.757	6.287
5	3.635	4.602	5.218	5.673
6	3.461	4.339	4.896	5.305
7	3.344	4.165	4.681	5.000
8	3.261	4.041	4.529	4.886
9	3.199	3.949	4.415	4.756
10	3.151	3.877	4.327	4.654
11	3.113	3.820	4.256	4.574
12	3.082	3.773	4.199	4.508
13	3.055	3.735	4.151	4.453
14	3.033	3.702	4.111	4.407
15	3.014	3.674	4.076	4.367
16	2.998	3.649	4.046	4.333
17	2.984	3.628	4.020	4.303
18	2.971	3.609	3.997	4.277
19	2.960	3.593	3.977	4.253
20	2.950	3.578	3.958	4.232
24	2.919	3.532	3.901	4.166
30	2.888	3.486	3.845	4.102
40	2.858	3.442	3.791	4.039
60	2.829	3.399	3.737	3.977
120	2.800	3.356	3.685	3.917
∞	2.772	3.314	3.633	3.858

Note: This table may be used in a posterior test following a one-way analysis of variance to compare more than two means, for both the balanced and unbalanced designs. In the table, p is equal to: (The difference in the ranks of a pair of treatment means +1).

Table C.10   Percentage Points of the χ^2 Distributions for an χ^2 Distribution With ν Degrees of Freedom. The Table Gives the Values of χ^2 Which are Exceeded with Probability α. See Fig. C.5

$\alpha =$	0.50	0.10	0.05	0.025	0.01	0.001
$\nu = 1$	0.45	2.71	3.84	5.02	6.64	10.8
2	1.39	4.61	5.99	7.38	9.21	13.8
3	2.37	6.25	7.82	9.35	11.3	16.3
4	3.36	7.78	9.49	11.1	13.3	18.5
5	4.35	9.24	11.1	12.8	15.1	20.5
6	5.35	10.6	12.6	14.5	16.8	22.5
7	6.35	12.0	14.1	16.0	18.5	24.3
8	7.34	13.4	15.5	17.5	20.1	26.1
9	8.34	14.7	16.9	19.0	21.7	27.9
10	9.34	16.0	18.3	20.5	23.2	29.6
12	11.3	18.5	21.0	23.3	26.2	32.9
15	14.3	22.3	25.0	27.5	30.6	37.7
20	19.3	28.4	31.4	34.2	37.6	45.3
24	23.3	33.2	36.4	39.4	43.0	51.2
30	29.3	40.3	43.8	47.0	50.9	59.7
40	39.3	51.8	55.8	59.3	63.7	73.4
60	59.3	74.4	79.1	83.3	88.4	99.6

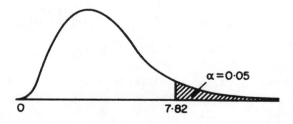

Figure C.5 χ^2 Distribution With $\nu = 3$ d.f.

Table C.11 Values of Pearson's *r*

| | Level of Significance for One-Sided H_1 | | | |
	0.05	0.025	0.01	0.005
	Level of Significance for Two-sided H_1			
d.f.	0.1	0.05	0.02	0.01
2	0.9000	0.9500	0.9800	0.9900
3	0.8054	0.8783	0.9343	0.9587
4	0.7293	0.8114	0.8822	0.9172
5	0.6694	0.7545	0.8329	0.8745
6	0.6215	0.7067	0.7887	0.8343
7	0.5822	0.6664	0.7498	0.7977
8	0.5494	0.6319	0.7155	0.7646
9	0.5214	0.6021	0.6851	0.7079
10	0.4973	0.5760	0.6581	0.7079
11	0.4762	0.5529	0.6339	0.6835
12	0.4575	0.5342	0.6120	0.6614
13	0.4409	0.5139	0.5923	0.6411
14	0.4259	0.4973	0.5742	0.6226
15	0.4124	0.4821	0.5577	0.6055
16	0.4000	0.4683	0.5425	0.5897
17	0.3887	0.4555	0.5285	0.5751
18	0.3783	0.4438	0.5155	0.5614
19	0.3687	0.4329	0.5034	0.5487
20	0.3598	0.4227	0.4921	0.5368
25	0.3233	0.3809	0.4451	0.4869
30	0.2960	0.3494	0.4093	0.4487
35	0.2746	0.3246	0.3819	0.4182
40	0.2573	0.3044	0.3578	0.3932
45	0.2428	0.2875	0.3384	0.3721
50	0.2306	0.2732	0.3218	0.3541
60	0.2108	0.2500	0.2948	0.3248
70	0.1954	0.2319	0.2737	0.3017
80	0.1829	0.2172	0.2565	0.2830
90	0.1726	0.2050	0.2422	0.2673
100	0.1638	0.1946	0.2301	0.2540

Table C.12   Values of Spearman's r_s

	Level of Significance for One-Sided H_1			
	0.05	0.025	0.01	0.005
	Level of Significance for Two-Sided H_1			
n	0.1	0.05	0.02	0.01
5	0.900	1.000	1.000	—
6	0.829	0.886	0.943	1.000
7	0.714	0.786	0.893	0.929
8	0.643	0.738	0.833	0.881
9	0.600	0.683	0.783	0.833
10	0.564	0.648	0.746	0.794
12	0.506	0.591	0.712	0.777
14	0.456	0.544	0.645	0.715
16	0.425	0.506	0.601	0.665
18	0.399	0.475	0.564	0.625
20	0.377	0.450	0.534	0.591
22	0.359	0.428	0.508	0.562
24	0.343	0.409	0.485	0.537
26	0.329	0.392	0.465	0.515
28	0.317	0.377	0.448	0.496
30	0.306	0.364	0.432	0.478

Table C.13　Coefficients for the Shapiro-Wilk Test for Normality

$n =$	2	3	4	5	6	7	8	9	10
a_1	0.7071	0.7071	0.6872	0.6646	0.6431	0.6233	0.6052	0.5888	0.5739
a_2	—	.0000	.1677	.2413	.2806	.3031	.3164	.3244	.3291
a_3	—	—	—	.0000	.0875	.1401	.1743	.1976	.2141
a_4	—	—	—	—	—	.0000	.0561	.0947	.1224
a_5	—	—	—	—	—	—	—	.0000	.0399

$n =$	11	12	13	14	15	16	17	18	19	20
a_1	0.5601	0.5475	0.5359	0.5251	0.5150	0.5056	0.4968	0.4886	0.4808	0.4734
a_2	.3315	.3325	.3325	.3318	.3306	.3290	.3273	.3253	.3232	.3211
a_3	.2260	.2347	.2412	.2460	.2495	.2521	.2540	.2553	.2561	.2565
a_4	.1429	.1586	.1707	.1802	.1878	.1939	.1988	.2027	.2059	.2085
a_5	.0695	.0922	.1099	.1240	.1353	.1447	.1524	.1587	.1641	.1686
a_6	0.0000	0.0303	0.0539	0.0727	0.0880	0.1005	0.1109	0.1197	0.1271	0.1334
a_7	—	—	.0000	.0240	.0433	.0593	.0725	.0837	.0932	.1013
a_8	—	—	—	—	.0000	.0196	.0359	.0496	.0612	.0711
a_9	—	—	—	—	—	—	.0000	.0163	.0303	.0422
a_{10}	—	—	—	—	—	—	—	—	.0000	.0140

$n =$	21	22	23	24	25	26	27	28	29	30
a_1	0.4643	0.4590	0.4542	0.4493	0.4450	0.4407	0.4366	0.4328	0.4291	0.4254
a_2	.3185	.3156	.3126	.3098	.3069	.3043	.3018	.2992	.2968	.2944
a_3	.2578	.2571	.2563	.2554	.2543	.2533	.2522	.2510	.2499	.2487
a_4	.2119	.2131	.2139	.2145	.2148	.2151	.2152	.2151	.2150	.2148
a_5	.1736	.1764	.1787	.1807	.1822	.1836	.1848	.1857	.1864	.1870
a_6	0.1399	0.1443	0.1480	0.1512	0.1539	0.1563	0.1584	0.1601	0.1616	0.1630
a_7	.1092	.1150	.1201	.1245	.1283	.1316	.1346	.1372	.1395	.1415
a_8	.0804	.0878	.0941	.0997	.1046	.1089	.1128	.1162	.1192	.1219
a_9	.0530	.0618	.0696	.0764	.0823	.0876	.0923	.0965	.1002	.1036
a_{10}	.0263	.0368	.0459	.0539	.0610	.0672	.0728	.0778	.0822	.0862
a_{11}	0.0000	0.0122	0.0228	0.0321	0.0403	0.0476	0.0540	0.0598	0.0650	0.0697
a_{12}	—	—	.0000	.0107	.0200	.0284	.0358	.0424	.0483	.0537
a_{13}	—	—	—	—	.0000	.0094	.0178	.0253	.0320	.0381
a_{14}	—	—	—	—	—	—	.0000	.0084	.0159	.0227
a_{15}	—	—	—	—	—	—	—	—	.0000	.0076

Table C.14 Percentage Points of W for the Shapiro–Wilk Test for Normality
Level of Significance

n	0.01	0.02	0.05	0.10	0.50	0.90	0.95	0.98	0.99
3	0.753	0.756	0.767	0.789	0.959	0.998	0.999	1.000	1.000
4	.687	.707	.748	.792	.935	.987	.992	.996	.997
5	.686	.715	.762	.806	.927	.979	.986	.991	.993
6	0.713	0.743	0.788	0.826	0.927	0.974	0.981	0.986	0.989
7	.730	.760	.803	.838	.928	.972	.979	.985	.988
8	.749	.778	.818	.851	.932	.972	.978	.984	.987
9	.764	.791	.829	.859	.935	.972	.978	.984	.986
10	.781	.806	.842	.869	.938	.972	.978	.983	.986
11	0.792	0.817	0.850	0.876	0.940	0.973	0.979	0.984	0.986
12	.805	.828	.859	.883	.943	.973	.979	.984	.986
13	.814	.837	.866	.889	.945	.974	.979	.984	.986
14	.825	.846	.874	.895	.947	.975	.980	.984	.986
15	.835	.855	.881	.901	.950	.975	.980	.984	.987
16	0.844	0.863	0.887	0.906	0.952	0.976	0.981	0.985	0.987
17	.851	.869	.892	.910	.954	.977	.981	.985	.987
18	.858	.874	.897	.914	.956	.978	.982	.986	.988
19	.863	.879	.901	.917	.957	.978	.982	.986	.988
20	.868	.884	.905	.920	.959	.979	.983	.986	.988
21	0.873	0.888	0.908	0.923	0.960	0.980	0.983	0.987	0.989
22	.878	.892	.911	.926	.961	.980	.984	.987	.989
23	.881	.895	.914	.928	.962	.981	.984	.987	.989
24	.884	.898	.916	.930	.963	.981	.984	.987	.989
25	.888	.901	.918	.931	.964	.981	.985	.988	.989
26	0.891	0.904	0.920	0.933	0.965	0.982	0.985	0.988	0.989
27	.894	.906	.923	.935	.965	.982	.985	.988	.990
28	.896	.908	.924	.936	.966	.982	.985	.988	.990
29	.898	.910	.926	.937	.966	.982	.985	.988	.990
30	.900	.912	.927	.939	.967	.983	.985	.988	.990

Acknowledgments

These Statistical tables are the same as in the 3rd edition (where the tables were in Appendix D), with the addition of Table C.9 and Table C.11. Table C.9 was adapted from a table in Pearson, E.S. and Hartley, H.O., *Biometrika Tables for Statisticians*, Vol. 1, 3rd edition, Cambridge University Press, Cambridge, 1966. Table C.11 can be derived from Table C.5 by using formula (14.2), namely:

$$Calc \ t \ = \ r \sqrt{\frac{n-2}{1-r^2}}$$

or, more usefully expressing r in terms of t,

$$r \ = \ \pm \frac{t}{\sqrt{t^2 + n - 2}}$$

where t is from Table C.5, for $\alpha = 0.025$ and $\nu = n - 2$.

Appendix D

Glossary of Symbols

The references in parentheses refer to the main chapter and section in which the symbol is introduced, defined, or used.

Roman Symbols

a	Intercept of sample regression line (15.2)
b	Slope of sample regression line (15.2)
B(n, p)	General binomial distribution (6.3)
c	Number of columns in contingency table (13.3)
d	Difference between pairs of values (9.10)
d	Difference between pairs of ranks (14.5)
$\bar{d}$	Sample mean difference (9.10, 10.13)
e	2.718 (2.4, 6.10)
E	Event (5.3)
E	Expected frequency (13.3, 16.2)
E′	Not E, complement of E (5.12)
G	Grand total of observed values in ANOVA (12.1)
H_0	Null hypothesis (10.1)
H_1	Alternative hypothesis (10.1)
m	Mean number of random events per unit time or space (6.10)
n	Number of observations in a sample (= sample size) (2.1, 9.1, 9.6, 9.9)

n	Total number of equally likely outcomes (5.3)
n	(Large) number of trials (5.3)
n	Number of Bernoulli trials in a binomial experiment (6.2)
n	Number of pairs (9.10, 10.13)
n	Number of individuals in correlation and regression (14.2, 15.2)
n!	Factorial n (2.2)
$\binom{n}{x}$	$\dfrac{n!}{x!(n-x)!}$ (6.3)
$N(\mu, \sigma^2)$	General normal distribution (7.3)
O	Observed frequency (13.3, 16.2)
p	Probability of success in a Bernoulli trial (6.2)
$P(E)$	Probability that event E will occur (5.3)
$p(E_1 \text{ and } E_2)$	Probability that both events E_1 and E_2 will occur (5.8)
$P(E_1 \text{ or } E_2 \text{ or both})$	Probability that either or both of events E_1 and E_2 will occur (5.8)
$P(E_2 \mid E_1)$	Probability that event E_2 will occur, given E_1 has already occurred (5.9)
$P(x)$	Probability of x successes in a Bernoulli trial or in a binomial experiment (6.2 and 6.3)
$P(x)$	Probability of x random events per unit time or space (6.10)
q	Studentised range statistic (Table C.9)
r	Number of trials resulting in event E (5.3)
r	Number of equally likely outcomes resulting in event E (5.3)
r	Number of rows in contingency table (13.3)
r	Sample value of Pearson's correlation coefficient (14.2)
r_s	Sample value of Spearman's correlation coefficient (14.5)
R_1, R_2	Sum of the ranks of values in samples of sizes n_1 and n_2 in Mann-Whitney U test (11.8)
s	Sample standard deviation (4.8)
s^2	Sample variance (4.12)
s^2	Pooled estimate of variance (9.11)
s_d	Sample standard deviation of differences (9.10, 10.13)
$s.e.$	Standard error in posterior test after ANOVA (12.5)
s_r	Estimate of residual standard deviation (15.6)

t	Student's statistic (9.5, 9.10, 10.9)
T	Wilcoxon signed rank statistic (11.6)
T_+, T_-	Sum of ranks of positive and negative differences in Wilcoxon signed rank test (11.6)
U	Mann-Whitney statistic (11.9)
U_1, U_2	Values calculated in Mann-Whitney U test (11.8)
x	Any variable (2.1, 6.2, 6.3, 13.2)
x	Variable used to predict y variable in regression (14.1)
$x_{(1)}, x_{(2)}, x_{(n)}, x_{(i)}$	The first, second, n^{th}, i^{th} values of x in order of magnitude (16.5)
x_0	A particular value of x in regression analysis (15.7)
$\bar{x}$	Sample mean of x (2.1, 4.2)
y	Variable used with variable x in correlation (14.2)
y	Variable to be predicted from x variable in regression analysis (15.1)
$\bar{y}$	Sample mean of y (15.1)
z	Value used in Table C.3 (areas of normal distribution) (7.3)

Greek Symbols

α (alpha)	Probability used in t tables (9.6)
α	Probability used in χ^2 tables (13.3)
α	Intercept of population regression line (15.6)
β (beta)	Slope (or gradient) of population line (15.6)
μ (mu)	Mean of a normal distribution (7.2)
μ	Population mean (8.5, 9.2, 10.3)
μ_d	Mean of a population of differences (9.10, 10.13)
$\mu_{\bar{x}}$	Mean of the distribution of $\bar{x}$ (8.5)
ν (nu)	Degrees of freedom (9.4, 9.7, 10.6, 10.13, 12.1, 13.3, 14.3)
ρ (rho)	Population value of Pearson's correlation coefficient (14.2)
σ (sigma)	Standard deviation of a normal distribution (7.2)
σ	Population standard deviation (8.5)
$\sigma_{\bar{x}}$	Standard deviation of the distribution of $\bar{x}$ (8.5)
Σ (sigma)	Operation of summing (2.1)
χ^2 (chi-squared)	Chi-squared statistic (13.1, 16.1)

Appendix E

Introduction to Minitab for Windows

In this appendix, there is a very brief introduction to Minitab for Windows. It is brief because nowadays it is reasonable to assume that you, the student, will have some experience in the use of 'spreadsheets' and/or 'Windows'.

However, if this doesn't apply to you, and you are starting 'from scratch', you are advised to study the following text first:
The Student Edition of Minitab for Windows by McKenzie, J., Shaefer, R. L., and Farber, E., ISBN-0-201-598886, published by Addison-Wesley-Longman, Reading, MA, 1995.
The Student edition will tell you all about the statistical computer package, Minitab for Windows, and what it can do; but it is essentially a computing textbook.

This book, *Essential Statistics* (ES4), is a Statistics textbook, in which Minitab is used as a tool to speed things up. In ES4 I have used the *Student Edition of Minitab* and the software which is provided with it to avoid problems which might arise if you (the student) and I used different versions.

The remainder of this appendix will demonstrate, using an example, how to do a number of basic procedures, and I assume that the *Student Edition of Minitab for Windows Software* has already been loaded into your computer:

Accessing Minitab—this may vary according to which version of Windows you are using.
Setting up a small database in a file.
Saving the file on the computer's main drive (C), and retrieving it at a later computing session.

Performing two statistical procedures on the data-base using the Minitab tools **Stat** and **Graph**.

Click on **Start**. Various menus appear. Move the pointer arrow to **Programs**, which is one of the items in the first column menu. Then move the pointer to the right into the second column and into the item **MINITAB Student**, and finally move the pointer into the third column and locate **MINITAB Student Edition**. Click once on this item. You should now see three windows, the one at the back has a label including the words 'Untitled Worksheet', the one in the middle is a 'Session' window, while the front one is a 'Data' window which has columns headed C1, C2, …, and rows labelled 1, 2, ….

Suppose we wish to set up a database containing the student data of Table 1.1 of this book. To keep things simple, we will choose the data for the first five students only, i.e., using only rows 1 to 5 in Table 1.1. Click in the box below C1 (and to the right of the arrow which points left to right) and ENTER (where 'enter' means 'type in using the keyboard') the name of the first variable, namely SEX, and press the enter button. Now enter the other variable names (HEIGHT, SIBS, DISTANCE, DEGREE, and ALCOUNT) and then enter the first five rows of data for columns C1 to C6. (Note that there is no need to enter the student reference numbers, since Minitab automatically numbers the rows.) You can move around the data window using the four arrow buttons next to the Ctrl button. Now you are ready to do some statistics, using the horizontal menu bar (which appears to be in the back window). So, Click on **File** (see back window) and then Click on **Save Worksheet As**. A new window should appear labelled 'Save Worksheet Window As.' Choose **Minitab Worksheet**, then Click on **Select File**. Yet another window appears. Make sure you have chosen C drive, then type in STDAT5.MTW in the box below **Filename**. The filename should have a maximum length of 12 letters including .MTW at the end. Click on **OK**. Choose **File** > **Display Data**, and enter C1-C6 in the box labelled **'Columns and Constants to display'**. Click on **OK**. The first five rows of data appear on the screen, and now is a good time to check for accuracy from Table 1.1. You have now saved the database for the five students in a file called STDAT5.MTW on the computer's C drive. If you went away and came back later, which is equivalent to Choose **File** and then **Restart Minitab**, you can retrieve your database by the following:

Choose **File** > **Open Worksheet** > **Select File** and 'enter' (meaning type in) STDAT5.MTW, then Click on **OK**. Try it!

So far, you will only have used **File** on the horizontal menu bar on the back window. We will now use **Stat** and **Graph**, as these are very useful in basic statistical methods. Here are two examples:

1. Choose **Stat** > **Basic Statistics** > **Descriptive Statistics** (Note the use of two > signs in the above. It is a kind of shorthand. For example, Choose Stat > Basic Statistics > Descriptive Statistics means select Stat from the horizontal menu, using keyboard or mouse, then point to Basic Statistics on a drop-down menu and then move the pointer to Descriptive Statistics on another drop-down menu and Click.). Returning to the example, on a new window you will see a large box labelled **Variables**. Enter C2-C4 (or C2 C3 C4), then Click on OK. You will see ten descriptive statistics for each of three variables. For example, the mean distance from home to Oxford is 105 km (more about the other statistics in Chapter 4 of this textbook).
2. Now try: Choose **Graph** > **Character Graphs** > **Dotplot**, and Click on the last of these. Again enter C2-C4 in the Variable box, and Click on **OK**. Each dot on each dotplot can be checked against the 'raw data' of Table 1.1.

Finally, to get a hard-copy printout of what we have covered in this 'session' (i.e., in this appendix), Choose **File** > **Print Window**. The output for this session should be the same as the following:

Table E.1   Data for Five Students from Table 1.1

Row	Sex	Height	Sibs	Distance	Degree	Alcount
1	1	183	1	80	2	6
2	2	163	2	3	1	32
3	2	152	2	90	1	22
4	2	157	3	272	2	12
5	2	157	1	80	2	12

Table E.2   Descriptive Statistics for Three of the Variables in Table E.1

	N	MEAN	MEDIAN	TRMEAN	STDEV	SEMEAN
Height	5	162.40	157.00	162.40	12.16	5.44
Sibs	5	1.80	2.00	1.800	0.837	0.374
Distance	5	105.0	80.0	105.0	99.7	44.6

	MIN	MAX	Q1	Q3		
Height	152.0	183.0	154.5	173.0		
Sibs	1.00	3.00	1.0	2.5		
Distance	3.0	272.0	41.5	181.0		

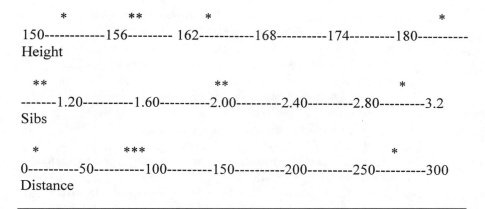

```
     *            **              *                              *
150------------156--------- 162-----------168----------174---------180----------
Height

  **                           **                          *
-------1.20----------1.60----------2.00---------2.40---------2.80---------3.2
Sibs

  *              ***                                        *
0----------50----------100---------150----------200--------250---------300
Distance
```

Figure E.1   Dotplots for Three of the Variables in Table E.1

Index